Praise for *Wisdom for a Livable Planet*

"Eye-opening profiles of some of the leading visionary environmentalists of our day. A pleasure to read and filled with wisdom. The book's insights will interest readers everywhere. It richly deserves a wide circulation."

—**Peter H. Raven**, George Engelmann Professor
of Botany, Washington University

"This book contains a number of home truths, calmly and moderately enunciated, that point the way toward a world more sturdy and robust than the troubled one we now inhabit. It is a kind of primer for twenty-first-century citizenship, and well worth the reading."

—**Bill McKibben**, author of *Wandering Home:
A Long Walk Across America's Most Hopeful Landscape,
Vermont's Champlain Valley and New York's Adirondacks*

"Despite rumors of 'The Death of Environmentalism,' *Wisdom for a Livable Planet* makes clear that green stalwarts are alive and well. McDaniel offers us fascinating accounts of eight living souls who continue to make vital contributions to an ecologically healthier world."

—**Max Oelschlaeger**, F. B. McAllister Endowed Chair
of Community, Culture, and Environment,
Northern Arizona University

"The visionaries McDaniel has chosen to highlight are pathbreakers in the endeavors humanity must take up if we and the rest of the biosphere are to survive—including wildlands conservation, ecological economics, ecological agriculture, and population reduction. Readers will find this an accessible and intelligent introduction to the most important conversations now taking place. McDaniel's own deep understanding of the formidable environmental challenges facing us informs every page of this clearly written, essential book."

—**Richard Heinberg**, author of *The Party's Over:
Oil, War, and the Fate of Industrial Societies*

Wisdom for a Livable Planet

The Visionary Work of

TERRI SWEARINGEN

DAVE FOREMAN

WES JACKSON

HELENA NORBERG-HODGE

WERNER FORNOS

HERMAN DALY

STEPHEN SCHNEIDER

AND DAVID ORR

Carl N. McDaniel

TRINITY UNIVERSITY PRESS

SAN ANTONIO

Published by Trinity University Press
San Antonio, Texas 78212

Copyright © 2005 by Carl N. McDaniel

Cover design by DJ Stout, Pentagram, Austin, Texas
Book design by BookMatters, Berkeley

♾ The paper used in this publication meets the
minimum requirements of the American National
Standard for Information Sciences—Permanence of
Paper for Printed Library Materials, ANSI Z39.48-
1992.

Library of Congress Cataloging-in-Publication Data

McDaniel, Carl N., 1942–

Wisdom for a livable planet : the visionary work
of Terri Swearingen, Dave Foreman, Wes Jackson,
Helena Norberg-Hodge, Werner Fornos,
Herman Daly, Stephen Schneider, and David Orr /
Carl N. McDaniel.

p.   cm.

Includes bibliographical references and index.

ISBN 1-59534-008-4 (alk. paper) —
ISBN 1-59534-009-2 (pbk. : alk. paper)
1. Environmental sciences. I. Title.

GE105.M385   2005
333.72--dc22   2004019081

09   08   07   06   05 / 5   4   3   2   1

for Mary, my lifetime partner

Contents

Acknowledgments *ix*

ONE Introduction *1*

TWO Clean Air: Terri Swearingen and Toxic Waste *5*

THREE Restoring Wildlands: Dave Foreman and
Preserving Biodiversity *34*

FOUR Healthy Farms: Wes Jackson and Agriculture *57*

FIVE Living Locally: Helena Norberg-Hodge
and Globalization *83*

SIX Be Fruitful and Few: Werner Fornos
and Population *112*

SEVEN Living in a Finite World: Herman Daly
and Economics *132*

EIGHT Accepting Uncertainty: Stephen Schneider
and Global Climate Change *164*

NINE Ecological Design: David Orr and Education *191*

TEN Can We Change, Will We Change? *216*

Notes *229*
Further Reading *261*
Index *267*

Acknowledgments

The heart of this book is the stories of visionaries who have generously allowed me into their lives so I could make the causes they have championed come alive by recounting their personal adventures of discovery. I write about eight people, each of whom, for the better part of his or her life, has taken on a fundamental challenge before us—Herman Daly (economics), Dave Foreman (biodiversity), Werner Fornos (population), Wes Jackson (agriculture), Helena Norberg-Hodge (globalization), David Orr (education), Stephen Schneider (climate change), and Terri Swearingen (pollution). I thank each of them not only for the time they so generously gave me but also for the trust and support they bestowed upon me and this project.

Writing for me is a team endeavor; family, friends, and colleagues are invaluable in transforming hackneyed drafts into understandable prose. David Benzing, Chris Bystroff, Jon Erickson, John Gowdy, David Hopkins, Jenny McDaniel, Virginia McDaniel, Margaret Stoner, Holly Sullivan, and John Wimbush commented insightfully on various chapters and provided useful suggestions. Doyle Daves, Kirk Jensen, Brad Lister, Stuart McDaniel, and Ian Sussex provided invaluable reviews of entire draft manuscripts. David Borton, Harriet Borton, Harry Roy, and Howard Stoner deserve special thanks for providing critical assessments of every chapter as it was written and Susan Blandy for proofreading. I thank Jean Thomson Black for valuable comments and for soliciting several reviews that enabled me to focus my message better. I am indebted to David Ehrenfeld who saw promise in an early draft and introduced me to Barbara Ras at Trinity University Press. Barbara passionately and creatively guided the production of *Wisdom for a Livable Planet*. With superb editorial skill Helen Whybrow and then Rachel Berchten masterfully polished what is before you. I am also most appreciative of the suggestions and support of Edward O. Wilson, who has been not only an inspiration to me but also a wellspring of ideas and conceptual understandings that have influenced my perspectives on the biological world.

Homo sapiens is the only animal that tells stories. Exactly when our ancestors evolved the capacity to talk is unknown, but humans have been talking for a long time—at least fifty thousand years and perhaps longer. Literacy, however, is another matter. Although absolutely essential in our technological civilization, it appeared as a cultural adaptation no more than a few hundred generations ago. Whereas preliterate societies relied on the collective memories of the tribe, modern humans have access to vast libraries of indexed and cross-referenced information that is free from the vagaries of forgetfulness. Yet, a storyteller is still constrained by the limited capacity of a single brain and led by the brain's uncontrollable propensity to weave connections and provide details that are, more often than we like to admit, fabricated to satisfy the brain's need to make sense of incomplete patterns.

The patterns I have seen and used to create the stories told here are unique and, of course, include a bit of my own unintentional fabrication. My stories are not, however, works of fiction but rather present reality as accurately as I can. The people mentioned above, and many others, have done their best to keep me honest and truthful, but I alone am responsible for what is written here.

ONE

Introduction

It takes me only five minutes to drive from our home in the city of Troy, New York, to open country of farms, rolling hills, and hardwood forests. The robust trees and lush growth I see on century-old abandoned potato farms indicate fertile soil and an abundance of life's blood—water. Such images make it seem more than reasonable to believe that we have resources and land to spread out on for a long time to come.

When I am hungry for a bagel, I go to Bruegger's Bagel shop downtown and have a selection of seventeen types of bagels with a choice of thirteen toppings. My wife, Mary, and I can see a movie at one of a dozen local theaters almost anytime. We can drive our car to visit friends and relatives or to go hiking in the nearby Adirondack, Catskill, or Green Mountains on a moment's notice.

Truly, life offers a cornucopia of endless choices for some of us in the industrialized world. Of course, living is a bit more crowded in places like New York City, London, or Tokyo, but there, too, choices abound. For the right price all imaginable types of entertainment, shops, and restaurants cater to every possible desire. This abundance tells part of me that "everything is just fine." Our culture gives us the message that a satisfying existence is only a question of money and, of course, of keeping the economy growing so that people have jobs and money to spend.

In spite of this good life, the availability of bagels with lots of toppings is not a particularly accurate predictor of the future. Notwithstanding the allure of the bountiful lifestyles we see, envisioning the future requires we look at the health of the fundamental biological and physical processes that underlie our successes, rather than the successes themselves. How can we do this when it is so easy to be blinded, and then seduced, by that which surrounds us?

I have spent the better part of my adult life struggling to resolve this question. On the one hand, as a recipient of the abounding wealth provided in the industrialized world, it seems that hard work, responsible planning, and opportunism—along with a bit of luck—are the basic ingredients to attain the good life. On the other hand, the major trends in population growth, biological diversity loss, climate change, degradation of resources like soil and water, and the long-term consequences of these trends leave no doubt that our current "good life" is tenuous. It is tenuous because we in Western culture have championed our successes and projected them into the future with little consideration of the fact that we are part of an interdependent biological enterprise whose principles and basic character will ultimately dictate success, and failure, for all life.

The Greek myth of the craftsman and inventor Daedalus and his son, Icarus, is a warning. King Minos imprisoned Daedalus and Icarus on the island of Crete in the labyrinth that Daedalus had designed. With escape by land and sea blocked, Daedalus fashioned wings of feathers and wax for both of them. Daedalus warned his son not to fly too high because the sun would melt the wax holding the wings together. As they flew from Crete, Icarus was enchanted with his new power. Ignoring Daedalus's advice and persistent warnings, Icarus flew higher and higher toward the hot sun. As he rose, the wax binding his wings melted, and Icarus fell to his death. The grieving Daedalus, whose advice had gone unheeded, flew on to Sicily.

Warnings are many and everywhere, yet we fly ever higher. The human population has increased from 1 billion to over 6 billion in the last two hundred years and is now globally 100 times denser than the population of any similar-sized animal in the history of Earth—truly an ecological anomaly. Humans use, directly or indirectly, about 25 percent of life's global energy flow, thereby impoverishing the rest of life. Ecosystems and species are being lost at a rate approaching that of the mass die off 65 million years ago when the dinosaurs went extinct. The carbon dioxide concentration in the atmosphere has increased by over 30 percent in the last century, and the decade of the 1990s was the hottest on record. Half of the world's forests have been cut down, while at least three-quarters of the major ocean fisheries are either fished out or in decline. Human activities use over half of the planet's readily available fresh water; many aquifers, such as the Ogallala aquifer under the Great Plains of the United States, are being mined with no possibility of recharge. Nitrogen and phosphorous have been the limiting nutrients in many ecosystems, but human activities in the last one hundred years have doubled the availability of these elements globally, causing chaotic perturbations of organismal relationships within numerous ecosystems. Human-created compounds like DDT, PCBs, dioxin, and furans are not only toxins but also hormone mimics that raise havoc in animal development and disrupt homeostasis, the balanced physiological state of a healthy organism. And these toxins are everywhere, even in distant and "pristine" places such as the Arctic. Rates of soil erosion in the United States are higher than they were during the dust bowl era of the 1930s, and the global loss of cropland to soil exhaustion, erosion, salinization, and waterlogging is 4 percent per decade. Well over half of the world's rangeland has been degraded. The stratospheric ozone layer continues to be reduced by chlorofluorocarbons and other human-generated compounds, resulting in increasing ultraviolet light at Earth's surface. And acid rain has increased substantially on all northern continents, threatening the health of forests and freshwater communities.

This list of environmental woes is familiar to anyone who reads the newspaper. Students of history know that cultural disintegration follows quickly on the heels of ecological collapse. Thus, despite our inability to predict with certainty specific outcomes, the broad patterns we see now do not bode well for global civilization. If the magnitude and character of the environmental changes experienced during the twentieth century are repeated in the twenty-first century, Earth's support systems will be overwhelmed, thereby radically impoverishing life and human existence.

While history enables us to perceive the overall consequences of how various peoples have lived, science unveils the underlying bases for these consequences. We are far from possessing a complete knowledge of the causes of all phenomena, but we know enough to provide for human well-being and to preserve a healthy planet for future generations. The question is: will we?

I don't know the answer, but I do know that many people have dedicated their lives to the belief that they can participate in creating just, ecologically centered patterns of living that are compatible with human nature. Thinking ecologically is not a passing fad or the venue of a special interest group; rather it is an emerging belief that all may share and that benefits everyone. It is a perspective that places us in appropriate relation with the rest of life. The ecological revolution is the next big idea in Western culture and has been in the making for more than a century. Religious, political, and economic freedom have been the big ideas that liberated Western culture, propelling it to become the dominant civilizing force of the past several centuries, but the successes of these big ideas have met the limits imposed by biological principles on a finite planet.

A small group of insightful people has perceived the biological and physical constraints now bearing down on humanity, and they have acted on their perceptions. I tell the stories of eight of them, each of whom has taken on a major environmental challenge that appears impossible to address effectively. Each story provides a window onto a different aspect of our environmental conundrum, while together the views from all these windows form a full picture of our unsustainable ways of life. At the same time, these visionaries' own lives, like innumerable others scattered around the world, are an inspiration. These men and women enable the rest of us to believe that answers to the challenges we face can be found. The narratives of these visionaries collectively give us hope, and their stories suggest to us ways to create a brighter future for all life.

Clean Air

Terri Swearingen and Toxic Waste

> On my return flight from San Francisco this week, I sat on the
> plane beside a young girl who told me she had just turned fourteen
> and was in San Francisco for vacation. She asked why I was in San
> Francisco, and I said I'd been at a gathering of environmental
> activists from all over the world to meet and discuss environmental
> problems. In response she gushed, "I just saw the movie *Erin
> Brockovich.* It was so good! Is that what you do?" At the invitation
> I explained that I was working to stop one of the world's largest
> hazardous waste incinerators from burning toxic waste next door
> to a four hundred–student elementary school. She immediately
> responded, "You don't have to do any research to know that's
> wrong!"
>
> TERRI SWEARINGEN

As Terri Swearingen and I walked out the front door of East Elementary,
we could see at eye level, between the houses across the street, the
vapor plume lazily drifting down river. On most days the plume
doesn't rise above the Ohio River bottomland but wallows in the valley, often drifting toward the houses and school on a bluff less than a
thousand feet away from its source, Waste Technologies Industries'
incinerator.

It was a gray spring afternoon with occasional drizzle. The scene
was like those seen in many working-class neighborhoods across the
industrial United States: children were riding bicycles around the
school and a baseball game was in progress on the athletic field;
the houses were in varying states of repair and the road needed resur-
facing. It was peaceful, and whatever sounds the people and cars made
were swallowed up by low-hanging clouds and misty rain.

Six months earlier I had met Terri Swearingen in San Rafael, California, at the Bioneers Conference where she began her presentation this way: "I'm a registered nurse, but my most important credential is that I'm a mother. Although I'm not a scientist and I don't have a Ph.D., I do have a few letters following my name." The image of her business card flashed on the screen as she read, "N.M.B.S.: No More Bull Shit!"

Swearingen lives in Chester, West Virginia, on the southeastern side of the Ohio River. A few miles away, in East Liverpool, Ohio, Waste Technologies Industries (WTI) operates one of the world's large toxic-waste incinerators. The incinerator sits on the northern floodplain of the river directly over two aquifers that are a backup drinking water source for several local towns. The East Liverpool area has at least 200 atmospheric inversions each year, during which an upper layer of stagnant warm air traps cooler air below so that the lower air mass is not dispersed. An infamous inversion not far from East Liverpool in 1948 led to the deaths of 18 people and made another 10,000 ill when for seven days industrial pollution was trapped over the vicinity of Donora, Pennsylvania, along the Monongahela River. About a third of the time, winds blow across the WTI facility directly toward East Elementary, which is just 1,100 feet from the incinerator's exhaust stack. The closest house is 320 feet from the facility, while hundreds more perch on bluffs or nestle in the surrounding valleys.

WTI's Resource Conservation and Recovery Act (RCRA) permit from the Environmental Protection Agency's (EPA) Region V office in Chicago states that WTI can burn a maximum of 179,000 tons of hazardous waste annually and can release, among other things, 100 tons of sulfur dioxide, 80 tons of particulates, 9,400 pounds of lead, and 2,560 pounds of mercury each year. In the late 1990s the plant burned about 70,000 tons of toxic waste annually, emitting undetermined tons of the authorized compounds in addition to unknowable quantities of other poisons like dioxins and furans. We know something about the symptoms of lead and mercury poisoning—weakness, anemia, paralysis, mental retardation, madness, death—and also that dioxins and furans can cause cancer, suppress the immune system, create developmental anomalies such as anatomical sex reversals by mimicking steroid hormones, and at higher levels can kill. We know little, however, about the effects on the health of plants and animals of hundreds of other compounds released from "state-of-the-art" incinerators, nor how these toxins move in ecosystems or to what extent they are found in our food.

As Swearingen drove me back to her home across the bridge connecting East Liverpool to Chester, I could see less than half a mile away the immaculate WTI facility. In the fading, late-afternoon light the vapor plume was invisible, as were the threats of its exhaust upon the life that clustered in this long-abused river valley. Yet, as Swearingen and her activist colleagues have known for more than twenty years, the WTI incinerator is likely to be relentlessly poisoning its neighbors, and any day its malfunction could result in a major accident.

At the beginning of the twentieth century the East Liverpool area touted itself as "the pottery capital of the world"—a huge teapot on a main street in Chester announces this history. At its height East Liverpool had 20,000 citizens; now about 13,000 people call it home. In the 1920s pottery was replaced by steel. Pottery was not a clean industry, but steel was even more devastating to the environment. While conservation and environmental preservation were on the minds of a few in the 1930s—such as wildlife biologist Aldo Leopold, who in 1935 purchased a worn-out Wisconsin farm and "shack" where he would write many of the essays that would greatly inform the coming environmental movement—the wider culture accepted pollution as a reasonable exchange, if not a prerequisite, for progress. We need only look at a map of the thousands of industrial toxic-waste sites across the United States, or in any other industrialized nation, to grasp the degree to which we dumped our wastes in the closest, most convenient places with apparently minimal thought to consequences.

The first half of the twentieth century was an exhilarating time as

we in the United States defended our principles and ourselves in two
world wars and built the most powerful nation on earth. Our technological prowess and industrial might gave rise to our belief that there
were no limits to making the material world conform to our desires.
By the 1970s, however, this naive view was no longer tenable, its
destructiveness exposed; the intoxicating effects of our successes were
giving way to more sober assessments. In my hometown of Troy, New
York, it was April 26, 1953, that marked the beginning of broadscale
awareness. The entire radioactive fallout cloud from Simon, a Nevada
atmospheric atomic bomb test on April 25, rained-out the next day on
the Troy area, giving residents a high dose of radiation—2.3 rads.
Within a decade Rachel Carson documented in her book *Silent Spring*
the ruinous consequences of other intentionally released toxins. The
fallout from the atomic bomb test Simon, along with Carson's writing,
brought to the public an awareness of the insidiousness of toxic materials. Nonetheless, since the fifties, with few exceptions, Americans
have been unprepared to stop using toxic materials or producing toxic
waste—or to manage both effectively.

In the second half of the twentieth century, the U.S. economy
would change. International competition combined with a host of factors to close the steel industry along the Ohio River, and by the end of
the 1970s the town of East Liverpool had three-quarters of a million
dollars in outstanding bills and unemployment of almost 20 percent.

Into this scene walked Don Brown, a 1964 East Liverpool High
School graduate and an employee of WTI in Arkansas. He returned
home to make the perfect match: East Liverpool would provide the
transportation infrastructure within a major industrialized area cov-
eted by WTI, and in return the city would receive the taxes and jobs
his company could provide. WTI was in the vanguard of the new rage
for incineration. The idea was to generate electricity by burning
industrial waste in a sophisticated incinerator with a destruction and
removal efficiency of 99.99 percent. This means that of the monitored
materials that go into the furnace only .01 percent would appear in the
stack gases released into the air. It sounded clean and safe to many,

including John Payne, a high school classmate of Brown's and, at the time, the mayor of East Liverpool. Construction would employ 200 people and, when operational, the incinerator would provide 100 permanent jobs. Property taxes of $1.25 million and city payroll taxes of $265,000 would increase by about a third the city's inadequate revenue stream. All the town had to do was to welcome this new facility to its community. It sounded too good to be true. And it was, because .01 percent of tens of thousands of pounds of noxious waste is a huge amount of toxic material to release into the air and subsequently to rain down on the land.

Terri Swearingen, a lifetime resident of the East Liverpool area, first learned about the planned incinerator in 1982 from a patient in the doctor's office where she was working. By that time several citizens' groups opposing the project had formed and she joined their efforts. Pregnant at the time, she recalls, "I knew this valley already had so much pollution. Respiratory diseases among children were so common here people thought it was part of childhood. And here I am, bringing a baby into the world." As a nurse and expectant mother, she could not understand how anyone could intentionally put tons of toxic compounds, including thousands of pounds of lead and mercury, into the air less than a quarter of a mile from an elementary school. Swearingen remembers, "[A]t the time I honestly believed in government. I thought the government would protect us."

As a nurse and health professional Swearingen automatically assumed that health trumped economics. She knew the government enforced strict standards for drinking water based on the number of coliform bacteria as well as on concentrations of poisonous materials like lead and mercury. She deeply believed in democracy and was certain that when government officials understood the health hazards posed by the incinerator and knew the people were unwilling to take these risks, they would not allow WTI to build the incinerator in East Liverpool, or perhaps anywhere. Surely, she thought, we can find ways not to produce all of these toxic wastes or at least to keep them from being released into the environment.

With the birth of her daughter in May 1982, Swearingen focused her energies on family and work, fully believing that the local opposition groups—Citizens Protection Association formed by local farmers and Save Our County formed by town residents—would prevail. However, for seven years the WTI project slowly, but persistently, overcame legal and regulatory challenges. In February 1983, the Ohio EPA issued a permit despite local citizens' major health and safety concerns about lead and mercury poisoning or a fire at WTI, as well as the unresolved liability issue of who was, in fact, responsible for damages if an accident did occur. Although WTI's application lacked the required signature of the owner of the land on which the incinerator was to be built, the EPA's Region V office granted the company its RCRA permit in June 1983. Challenges delayed the effective date of the ten-year RCRA permit until January 1985. The Ohio Hazardous Waste Facilities Approval Board overrode the negative recommendation of its sole hearing examiner, Richard Brudzinski, and granted WTI a permit in April 1984. In August of that year Ohio established new siting criteria stating that hazardous waste facilities could not be on floodplains and must be at least 2,000 feet from schools, hospitals, prisons, or residences. However, these new siting criteria were never imposed on WTI. Opponents raised numerous objections to the various permits granted to the company, but no permits were withdrawn, even though government agencies were given reasons and opportunities to do so.

In 1989 Swearingen noticed that a substantial part of a hill in West Virginia was being loaded onto trucks and taken across the river to the incinerator site. Site preparation had begun! Her biggest fear at the time was that there would be an accident at the plant and people in the community would be hurt or killed before anything could be done. "I realized that this [incinerator] would become a reality unless we stopped it," she recalled, "and I made up my mind if I was not ready to make a commitment, how could I expect anybody else to stop it?"

In the tradition of Lois Gibbs from Love Canal in New York and Anne Anderson from Woburn, Massachusetts (where cancer clusters were linked to contaminated well water), Terri Swearingen was to become an environmental activist of the first rank. These women, like Swearingen, were mothers whose activism was a response to seeing their children become victims of toxic insults. Gibbs's son had suffered from convulsions and severe asthma after attending the grade school built on Love Canal, a trench filled with toxic wastes by Hooker Chemicals and the city of Niagara Falls. Anderson's son had died of leukemia apparently caused by well water chemically contaminated by a Beatrice Foods' tannery and a facility of W. R. Grace that made machinery and used trichloroethylene, a carcinogen, for cleaning small metal parts. As with Gibbs, Anderson, and a host of other activists—mostly women—the threat to the community, and to children in particular, brought forth in Swearingen an intensity of commitment unlike any other she had previously experienced.

Looking across the Swearingen hilltop property I saw open grassland with trees and hills in the distance. It is the type of location many people would choose for their home, if they could. The serenity evoked by this idyllic setting contrasted sharply with that of the toxic waste–contaminated floodplain and the WTI site to which yet another West Virginia hill had been moved to raise the land above the hundred-year flood level. Neither Swearingen's home nor her seven-year-old daughter, Jaime, were as directly threatened by the incinerator as were the children at East Elementary. Yet seeing the actual construction at that location elicited from Swearingen an emphatic response—she was going to stop WTI.

Not knowing how she was going to stop it, she began by reading books, among them, *Rush to Burn: Solving America's Garbage Crisis?* She then proceeded to call the activists and scientists mentioned there, who in turn recommended more readings. Soon papers, books, articles, and correspondence filled the tables, chairs, and floor of her home, claiming all her time. Swearingen told me that she once took an incinerator book to read during a football game, concluding matter-of-factly, "It consumed my life."

Swearingen had known from the beginning that locating an industrial toxic-waste incinerator, originally described as the "world's largest," less than a quarter of a mile from an elementary school was a bad idea. Just how bad was convincingly demonstrated in October 1990 when construction workers on the WTI site unintentionally broke a gas main. In no time, because of its closeness to WTI, East Elementary was engulfed in gas. School personnel, fearing a broken gas line at the school, evacuated the children. Many had headaches and nausea, a few vomited. If any volatile toxins were released in an accident at WTI, they would engulf East Elementary.

The construction of the WTI incinerator had clearly posed both risks and dangers to the schoolchildren and residents close by, but new information raised questions about the safety of the entire process of toxic incineration itself. In the late 1970s, the idea of WTI's incinerator had appealed to Mayor Payne and to many residents in the area, not only for financial reasons, but also because the technology was said to be capable of burning toxic waste so thoroughly that the destruction and removal efficiency was 99.99 percent. To industrial communities like East Liverpool with a hundred-year history of dangerous, dirty, polluting industries, removal of 99.99 percent sounded like virgin purity. But by the early 1990s, many incineration experts and environmentalists like Swearingen who had done their research knew incineration was a bad way to handle waste. The emissions were only part of the waste. In addition to residual hazardous-waste liquids—9 percent of original volume—the tons of ash that remain after incineration—29 percent of the original volume—are more toxic than the initial waste and require disposal in leach-proof landfills. From the beginning the process requires careful and persistent monitoring, and although new equipment may work well at first, problems arise soon after operation is begun. Studies show that often incinerators do not work as advertised and that monitoring is frequently ineffective in preventing accidents.

Unfortunately, because of the focus on incineration for handling waste in the 1980s, the debate in the 1990s had focused on "How effectively can 'state-of-the-art' incineration eliminate toxic waste?" and not on the pertinent question, "Is incineration the best way to deal with toxic waste?" or on the even more fundamental question, "Can we stop generating toxic waste?" Since the siting permits had been granted and construction had begun, to stall the incinerator's operation opponents of WTI had to question its efficiency and safety while at the same time forcing a challenge to the underlying assumptions about the process into the public discussion.

In the tri-state area (Ohio, West Virginia, Pennsylvania), by the early nineties, more than ten years of legal, regulatory, and political turmoil had given birth to numerous citizen and business groups on both sides of the debate. Swearingen quickly became a formidable presence among the grassroots opponents and took the leadership role in creating the Tri-State Environmental Council (T-SEC) that became the clearinghouse for opposition groups including Save Our Community, Save Our State, Stewartville-Weirton Area Against Incineration, Let's Improve Valley Environment, Southwestern Pittsburgh Pennsylvania Alliance for a Clean Environment, and Pittsburgh Against Toxic Incineration.

Informational meetings, regulatory hearings, court challenges, and ultimately civil disobedience became T-SEC's means for drawing attention to the hazards of WTI's incinerator and to the potential dangers of its location. The group also raised awareness about more fundamental issues, such as reducing the production of toxic waste and finding better ways to prevent industrial poisons from moving freely in the environment.

The T-SEC activists quickly learned to expose the inner workings of a tragically flawed decision-making process wherein the wrong questions were often those considered while the larger, more important ones were not discussed or open to debate. A good example of T-SEC's getting the right questions on the table was in the fall of 1990 when WTI requested permission to alter its technology by adding a spray-dryer system to reduce the volume of liquid waste. This request enabled T-SEC to ask for an Ohio EPA hearing. East Liverpool

Middle School's auditorium was crowded with 350 people on January 17, 1991, when EPA staff attorney Karen Haight opened the hearing. When Haight called Michael Stein from Save Our State, Stein yielded his time to Paul Connett, a chemistry professor and incineration expert from St. Lawrence University. Speaker after speaker yielded time to Connett or to one of the other experts T-SEC had brought to the hearing.

Experts opposed to WTI dominated the hours-long hearing despite Haight's attempts to stop them. When chided by Haight, Connett replied, "I've come a long way and I've got a lot of information that these people need to hear, and I don't think you are going to stop me." Connett explained how the spray-dryer technology would increase the chlorinated organic material, dioxins, lead, and other toxins in the exhaust, not just water vapor as stated in WTI literature. He noted that federal regulations allow the original permit to be reconsidered if new information indicates "a threat to human health or to the environment exists, which was unknown at the time of permit issuance."

Hugh Kaufman, an expert witness brought in by T-SEC from US EPA's hazardous waste division, observed that since detailed technical information on the spray-dryer technology was confidential, "the public does not have the opportunity to have technical experts to review the technical information and give you adequate responses. . . . I don't see how you can go ahead with the regulatory process, unless this is a sham." Michael McCawley, a civil engineering professor at West Virginia University (WVU), said the review process lacked a safety analysis so engineering design mistakes would go undetected, and Robert Diener, a waste management professor at WVU, concluded that incineration is "the wrong way to dispose of waste." Connett summed up with, "Nobody, nobody, in 1991 would sanction the building of a hazardous-waste incinerator here. . . . You have to find a way of denying this permit." The Ohio EPA didn't even open discussions on the permit, but Swearingen and the T-SEC had changed the dynamics of the protest. They could not control the agenda or determine the outcome, but by creatively voicing opposition they had made it a dramatically altered engagement.

George V. Voinovich, a former mayor of Cleveland and currently a U.S. senator from Ohio, had been elected Ohio's governor in fall 1990. Environmentalists in general and WTI opponents in particular considered this bad news. Whatever sympathy the Ohio EPA had for the protesters' position slipped away as Voinovich's appointees filled key positions. As spring came, Voinovich's stance was clear: he would not interfere with the completion of the WTI facility. With the first incinerator under construction, WTI secured contracts for eliminating more waste than the one incinerator could handle. To achieve the capacity to burn the authorized annual 179,000 tons of waste, the company initiated plans to build a second incinerator at the East Liverpool site, but these would never be realized.

Swearingen knew that something had to be done to overcome the substantial momentum that the advocates for WTI had gained. On May 5, 1991, an opposition rally of 1,500 from the three states surrounding the Ohio River joined "hands across the river." At the event Swearingen met Vincent Eirene, who knew Martin Sheen, the actor-activist who opposed nuclear arms and advocated on behalf of homeless people. Eirene told Sheen about the planned WTI incinerator, and within days the actor was on board. On June 2 Sheen toured the area with stops in Pittsburgh and Beaver, Pennsylvania; East Liverpool, Ohio; and Wheeling, West Virginia. There he was joined by West Virginia's governor and attorney general in opposition to WTI. Sheen's participation catapulted East Liverpool's plight into the national spotlight.

Sheen was a key speaker at rallies in June inaugurating a Greenpeace-sponsored two-week march on Columbus to convince Governor Voinovich to change his position. Marchers started from Steubenville, Nova, and Dayton and came from 70 to 130 miles to converge at Columbus on June 29. The marchers demanded legislation to reduce the use of toxic chemicals in industry and a moratorium on any increased incineration. The WTI protest had gone statewide and beyond. The aggressive, highly visible campaign gained support across the tri-state region as West Virginia's attorney general, the Ohio State Medical Association, the Pittsburgh City Council, the Allegheny County Board of Commissioners, and other groups stepped forward to denounce WTI plans and to make various proposals and resolutions. WTI responded with financial donations apparently intended to appease and influence these opposition groups, but these contributions led to even more controversy.

In September 1991, protesters had a meeting with Governor Voinovich. He said that the incinerator was safe, all of the appropriate permits had been granted, and he could do nothing to prevent its operation. He offered to commission a study of the health of nearby residents before and after WTI began operation to see if any resident's health had been adversely affected. Swearingen's activism had not only made her knowledgeable about WTI's planned operations and the possible effects in her area, but more important she had also come to grasp the wider implications of burning toxic waste: because incineration does not make toxic waste disappear but rather concentrates and disperses it into the environment where it moves freely, the planned incinerator would pass on the burden of toxic insult to people and places within and outside the tri-state area. In response to the governor's offer, Swearingen denounced his proposal as an unethical, badly designed experiment that would make guinea pigs of human beings. In another response, Vincent Eirene, from Pittsburgh, concluded, "There's nothing left but massive civil disobedience."

On October 13, 1991, the acts of civil disobedience began with a rally in downtown East Liverpool, where about a thousand protesters were fired-up by Paul Connett and Martin Sheen. The actor demonstrated the strength of his commitment to oppose WTI's plans when he told the crowd, "I will go all the way with you. You all know what I do for a living. This is what I do to stay alive."

Among the emotionally charged protesters were a core group. A dozen years of hearings, meetings, protests, legal challenges, apparent violations of law by government agencies, and misleading statements by industry and government officials had motiviated Sheen and these 32 concerned citizens—mothers, teachers, grandparents, politicians, health professionals, spouses of elected officials, college students, and retired steelworkers—to pledge to do "whatever it takes" to prevent the incinerator from starting operation.

The crowd moved from downtown to the incinerator site. Emotions ran high as Sheen announced, "I feel it's necessary to go over the gate now." He did not ask anyone to follow him as he scaled a seven-foot-high side gate. Once on the other side he knelt and prayed aloud in WTI mud. It was a tense moment for Swearingen and the other 31 protesters on the public side of the gate. Until this event most had never contemplated such an overt violation of civil conduct; risking arrest had been unthinkable. But one by one, encouraged by their supporters and recorded by the host of media present, all 32 went over the fence to be arrested and taken to jail. They came to be known in the media as the East Liverpool, Ohio 33—ELO33.

The ELO33 had trespassed and would be tried for that offense, but in their view the real lawbreakers were those who had allowed the WTI travesty to continue for so long. The ELO33 went to Columbus 11 days after being arrested to put out a "WANTED" notice for Governor George Voinovich. The posters appeared across Ohio, describing the governor as follows:

EYES: BLIND TO THE FACTS.

EARS: DEAF TO THE CALLS OF CITIZENS.

HEIGHT: SHRINKING UNDER PRESSURE FROM POLLUTERS AND THOSE WANTING TO TURN OHIO INTO THE WORLD'S WASTE CAPITAL.

DISGUISE: SOMETIMES PRESENTS HIMSELF AS A PUBLIC SERVANT.

LAST SEEN: BACKING AWAY FROM PROTECTING THE CITIZENS OF EAST LIVERPOOL FROM THE WORLD'S LARGEST HAZARDOUS WASTE INCINERATOR.

CAUTION: THIS MAN IS A DANGER TO THE PEOPLE OF OHIO!

Swearingen's humor, which had been so much a part of her family life, took on a biting, satirical character and began to appear in the campaign. A week after the WANTED poster for the real criminal came out, Swearingen and about two dozen colleagues went to Governor Voinovich's mansion to post FOR SALE signs indicating that he had sold out the citizens of Ohio for toxic waste. Just as they pulled up in their yellow school bus, Voinovich left the house headed for his car. From the sidewalk Swearingen yelled, "Governor Voinovich, we want to talk to you. You're misinformed." As he climbed into the backseat, he yelled back, "I'm not going to talk with you. I have nothing to say." Frustrated, Swearingen quietly said, "He's a weenie."

The next morning Cleveland's *Plain Dealer* reported the exchange, highlighting her "weenie" comment. At first she was embarrassed by this publicity, but the remark would soon be turned to her advantage in acts of political theater staged by the anti-incinerator coalition. In late November with Greenpeace backing, protesters brought barbecue grills, tofu dogs, hot dogs, and all the fixings to curbside in front of the governor's mansion. At the picnic Swearingen in a hot dog costume and Voinovich face mask playfully impersonated the governor, saying, "Frankly, I don't relish the pickle I'm in. I need to mustard the courage to ketchup with front-end prevention. I feel like a weenie being squeezed between both sides of the bun: one side represents citizens who don't want to be poisoned, and the other side's the polluter, who wants the profit. It's no picnic being in this position. I'm gonna get roasted."

Constant publicity, harassment, and documentary evidence—all brought to the governor by T-SEC and its backers—motivated Voinovich to call it quits on December 4, 1991, saying, "Enough is enough." He would push the legislature to enact a moratorium on the construction of any more toxic waste incinerators. T-SEC's analysis of government and industrial data showed that Ohio, even before WTI, already had sufficient capacity to burn all of its toxic waste and more. Voinovich admitted that, if he had had all the facts, he might have opposed WTI, concluding, "One might argue that we might not be taking this action now had it not been for the protests." Ironically, while declaring his opposition to incineration by requesting a moratorium, he still maintained, "It's a safe site."

This was a major victory for those opposed to WTI's proposed incinerator! Other communities in Ohio would not have to fight against the construction of toxic waste incinerators. At the same time, however, WTI construction continued, so the anti-incinerator campaign persisted. Thousands of packages of hot dogs and buns in Columbus grocery stores bore an extra label reading SEND HIM A WIENIE; CALL HIM AT (614) 466-3555 and picturing a hot dog in a bun on a plate. Below the plate was the message: GOVERNOR VOINOVICH IS A WIENIE ON WASTE; TELL THE GOVERNOR: "DON'T BURN OHIO, STOP INCINERATION." Every time Swearingen and her colleagues were in Columbus they ordered a hot dog and had it delivered to Voinovich's office.

On December 10 protesters waving foil-covered hot dogs brought to the governor's office a huge coloring book showing the simple math of toxic waste capacity. According to the government's own data, the two proposed incinerators in East Liverpool would be able to burn three times the amount of toxic waste that Ohio's industries were projected to generate in 1995. Furthermore, Ohio's existing commercial hazardous waste incinerator already handled 86,400 tons, which was 30 percent more toxic waste than Ohio was expected to produce in 1995. Clearly, these statistics demonstrated that the two additional incinerators were unnecessary.

A week following the math lesson on incineration capacity, about 50 people took over the Ohio EPA offices. Twelve burst into Director Donald Schregardus's office and, in his absence, handcuffed themselves together around his desk while the others roamed the halls handing out "pink slips" to employees for failing to protect Ohio's environment. The protesters demanded three things: the release of all files on WTI, an immediate halt to processing the revised WTI permit that allowed it to include the spray-dryer technology, and a commitment to require Ohio industry to reduce the use of toxic chemicals by half in five years. It was early evening by the time authorities had cut the handcuffs from the protesters surrounding the EPA director's desk.

Along with the other nine who had refused to leave, Swearingen was arrested. The next day the Ohio EPA issued WTI its revised permit.

The ELO33 went on trial in February 1992. As had been shown in *Anne Anderson, et al. v. W. R. Grace & Co., et al.* (a lawsuit brought as a citizen's response to the cancer cluster in Woburn, Massachusetts)— and many other environmental lawsuits—going to trial does not ensure justice. The legal and scientific approaches to truth are quite different. Science, although at times contentious, is a cooperative community affair where all parties work together to seek explanations and truth, whereas our legal system is an adversarial one in which each party argues for its self-interest. Few would now contend that justice was done for the cancer victims in Woburn, because in that case the plaintiff's legal representation, hampered by the legal system, was prevented from submitting scientific evidence in support of Anderson's case. The Anderson suit demonstrated that the outcome of a legal case depends a lot upon what evidence and testimony the judge allows and doesn't allow. In the ELO33 trial, Judge Melissa Byers-Emmerling permitted the defendants to argue that they resorted to civil disobedience because their illegal trespass sought to prevent greater harm, a necessity defense. With this ruling, Judge Byers-Emmerling, unlike Judge Walter J. Skinner in the Woburn case, not only leveled the playing field for the environmentalists, but also opened her courtroom for a televised debate on the science of incinerating toxic waste. One expert witness after another presented evidence of the potential health consequences of incinerating toxic waste, the inappropriateness of the site—especially the facility's proximity to a school—and the government-industry collusion. When Hugh Kaufman from the US EPA was asked if he thought the defendants could have done anything other than acting in civil disobedience, his response affirmed the necessity of their actions and cast doubt on the integrity of the state and federal EPA: "I don't, because the federal and state EPA have violated a number of statutes. To appeal to decision-makers who have chosen to ignore the law after a while becomes meaningless."

The eight jurors took two and a half hours to find all defendants not guilty. For the first time in Ohio, an acquittal resulted from the necessity defense. It was a noteworthy victory for the environmentalists' public relations effort, because the WTI incinerator had become a major national issue in which the serious downsides of toxic waste incineration were exposed. The plans for the facility, however, had not changed, and East Liverpool's incinerator would soon be ready for its test burn. The federal EPA had to authorize the burn, which would establish that the facility met the required destruction and removal efficiency of 99.99 percent for the measured chemicals. A satisfactory trial burn would enable the incinerator owners to be granted an operating permit.

Spring, then summer, passed without the test burn. It was held at bay by protests, civil disobedience, legal challenges, a hunger strike, congressional testimony, letters from prominent scientists, revelation of more government-industry collusion, efforts of numerous people in high places, and publicity like an ABC *Nightline* debate, in which Swearingen overwhelmed Region V's EPA administrator with her detailed knowledge of incineration and WTI's history.

The presidential race was getting under way in the summer of 1992 when vice presidential candidate Al Gore at a campaign stop in Weirton, West Virginia, signed a NO WTI poster and declared, "I'll tell you this, a Clinton-Gore administration is going to give you an environmental presidency to deal with these problems. We'll be on your side for a change." WTI opponents took note, but they understood that often candidates say things they can't deliver later.

During the third week of November numerous acts of civil disobedience occurred at the WTI facility, and over five days many protesters were arrested, as follows: Monday, 8 grandparents; Tuesday, 11 parents; Wednesday, 5 medical professionals; Thursday, 11 union members; Friday, 6 small business owners. And on Sunday, 500 people attended a rally at WTI and 75 were arrested.

What would it take to stop the WTI juggernaut that had overcome large-scale opposition for more than a decade? On December 7 vice

president–elect Gore addressed the Clinton-Gore administration's
first environmental issue with a promise to the East Liverpool community: "Serious questions concerning the safety of an East Liverpool,
Ohio, hazardous waste incinerator must be answered before the plant
may begin operation. The new Clinton-Gore administration . . . [will]
not issue the plant a test burn permit until . . . all questions concerning the compliance with state and federal law have been answered."
Excitement swept over Swearingen and her weary collaborators.
Perhaps victory was at hand.

Al Gore's book *Earth in the Balance: Ecology and the Human Spirit,*
published in 1992, sported this dustjacket comment of the late astrophysicist Carl Sagan: "A global environmental crisis threatens to over-
whelm our children's generation. Mitigating the crisis will require a
planetary perspective, long-term thinking, political courage and savvy,
eloquence and leadership—all of which are in evidence in Al Gore's
landmark book." Since its publication, several hundred thousand
copies have been sold and most people concerned about our environmental crises have heard of, if not read, Vice President Gore's book.
But, even for an American vice president, having written a book that
advocates an ecological perspective does not ensure implementation of
environmentally sound policy.

Could it be that environmentalists had expected too much from a
vice president who as a senator argued for meaningful resolution of
our environmental crises? No sooner had Gore proclaimed support for
the principles stated in *Earth in the Balance* did opponents explain in
no uncertain terms via a host of venues, including the editorial pages
of the *Wall Street Journal,* the political price of lost support from the
business community that would be paid for such policies. Our culture
tolerates environmentalism only so long as it has minimal impact on
big business. Obstruct the highway of commerce and you are roadkill.
At the time, this economically centered worldview deeply divided
America's citizens, even in East Liverpool—and it still does. Many of
the town's residents wanted the WTI facility and the jobs it would provide; and they wanted Terri Swearingen to leave. In Bill Clinton's 1992

campaign headquarters, someone posted a slogan that expresses the United States' highest priority: "It's the economy, stupid!" In an economically centered culture, jobs come first, not the health of people or the environment.

Clinton and Gore almost immediately backed off from their commitment to block the test burn until a detailed review had been conducted. The outgoing Bush EPA administrator, William Reilly, offered in early January 1993 to delay approval of the WTI test burn, thereby leaving it to the new administration to handle. He explained the response to his offer this way: "[Katie McGinty of the Clinton-Gore transition team] said the Vice President elect had had second thoughts about his position, had concluded that he should not interfere in the regulatory process and that the transition team would be grateful, the Vice President elect would be grateful, if I simply made that decision before leaving office." So on January 8 the Bush EPA gave WTI approval to conduct its test burn.

Immediately, Greenpeace, representing local citizens, sued to prevent the burn. US District Court Judge Ann Aldrich in Cleveland granted a halt of the test burn pending the outcome of the Greenpeace suit. The trial ended on March 5 and the court order allowed the test burn but stated that the incinerator could not operate until the US EPA gave final permission, a decision that would not be made for about another year. The evidence presented had convinced Judge Aldrich that if the incinerator were operational while the EPA deliberated on the permit, it "may pose an imminent and substantial endangerment to health and the environment." This was an unacceptable ruling for WTI, because the company calculated a cost of $150,000 for each day during which it did not operate. WTI quickly appealed Judge Aldrich's ruling in the US Court of Appeals for the Sixth Circuit in Cincinnati, arguing that the Cleveland district court did not have subject matter jurisdiction over the case and therefore could not prevent WTI from operating after a satisfactory test burn. The appeals court decided that once state and federal agencies were satisfied with the results of the test burn, WTI could operate while the court heard the case concerning the district court's jurisdiction.

WTI conducted its test burn in March and failed to meet required destruction and removal efficiency for three compounds: carbon tetrachloride, mercury, and polychlorinated dibenzodioxins. The permitting agencies considered these failures fixable or at least not serious enough to prevent operation. In April the EPA allowed WTI to commence limited burning, in advance of the final EPA decision to be made in about a year. Incineration began. WTI had prevailed.

If human and environmental health are primary objectives in a civilized society, then the WTI facility should not have been built in the East Liverpool location on the Ohio River in the first place, and it certainly should not have become operational. Both, however, came to pass. In 1980 when East Liverpool citizens Rebecca Tobin and Alonzo Spencer organized the early resistance, Swearingen believed in democracy and that her government would put the health and well-being of its citizens before economic interests. But as the outcome of the struggle against building and operating the facility demonstrates, Swearingen found that in this case our existing institutions proved ineffective in implementing appropriate health-based policies in a complex environmental case. In protest of this failure Swearingen often wears a button with the phrase WHATEVER IT TAKES over a red circle that contains the letters WTI with a red line through them.

While citizen pressure had brought about a moratorium on the building of more toxic waste incinerators in Ohio, it had not prevented WTI from moving toward full commercial operation. To prevent the facility from becoming fully operational, Swearingen and her colleagues decided to intensify the pressure on the new administration in Washington. The T-SEC invited newly elected Bill Clinton and Al Gore to East Liverpool. They wouldn't come, so Swearingen and seven colleagues went to the White House on March 18. With a thousand other tourists the eight activists saw the sights, but on entering the

Grand Ballroom, they formed a circle and sat down. As they sang "We shall stop the burn" to the tune of "We Shall Overcome," astonished tourists looked on as Secret Service agents took charge and the tour ended. For the sixth time Swearingen was taken to jail, on this occasion for "failure to quit." The arrest took place on Thursday. On the next Saturday about a hundred people protested in Lafayette Park across from the White House with a new slogan: FAILURE TO QUIT—GUILTY AS CHARGED.

A month later with a Pinocchio nose on her face fashioned from a toothbrush-holder and a WTI SPELLS DEATH poster under her arm, Swearingen took President Clinton's hand as he waded into the crowd at Pittsburgh International Airport after a speech on the economy. He tried to pull away, but she continued to hold his hand and asked him to fulfill his campaign promise to the citizens of East Liverpool. According to Swearingen, he answered, "We just met on this yesterday. We're not sure if there's anything we can do, because it was approved under the Bush administration." She then informed Clinton that she had a plan of what to do and that she wanted to meet with him. He did not respond. Other protesters were strategically located, and two more had words with Clinton. Joy Allison was told the same thing as Swearingen, but Alonzo Spencer reported that the president said, "I will try."

In mid April 1993, Swearingen joined a four-week Greenpeace tour of 25 communities in 18 states affected by toxic chemical waste. The tour ended on Pennsylvania Avenue right in front of the White House, where the group parked a mock incinerator spewing black clouds of nontoxic emissions. The protesters shouted in unison: "Al Gore, read your book. If you can't stop WTI, how can you save the planet?" A number of them handcuffed themselves to the "incinerator" stack and to concrete columns inside the van so that traffic was snarled for about six hours. As the police jackhammered the smokestack, demonstrators chanted, "Right tools, wrong stack." Swearingen and 53 colleagues went to jail. The next day Carol Browner, head administrator of the EPA, announced a freeze on new commercial hazardous-waste incinerators as well as a licensing review of those incinerators now operating on interim status. Nothing, however, was changed for the WTI facility.

In mid July, T-SEC brought about two dozen students from East Liverpool to see President Clinton, but he was out of town. The bus then went to the Swiss embassy (a U.S. subsidiary of the Swiss company Von Roll A.G. owned the WTI incinerator in East Liverpool), where the children were discharged before 23 demonstrators hand-cuffed themselves to the bus in the middle of the street. It took police several hours to cut the protesters free. Again Swearingen and her colleagues were arrested and taken away.

When President Clinton gave a speech on the front steps of the West Virginia capital in mid-August, the following day newspapers in the tri-state region pictured Clinton at the podium with Swearingen in front of him, hoisted on somebody's shoulders, holding a poster with the question WHAT ABOUT WTI? written over a skull and crossbones.

For the purpose of staging another dramatic protest, Swearingen and her T-SEC colleagues requested that people send their copies of *Earth in the Balance* to them. On November 6 about three hundred people from all over the country rallied in Lafayette Park across from the White House to "throw the book at Al." They initiated their national book recall by marching across the street to give a hundred copies of *Earth in the Balance* back to Gore.

While the demonstrations kept the issue in the public eye and on the administration's mind, the WTI incinerator burned toxic waste, and the court proceedings, permit challenges, and EPA reviews continued. On November 19 the US Court of Appeals for the Sixth Circuit in Cincinnati gave its final ruling, concluding that Judge Aldrich's court could not block the facility's operation. According to the ruling, the Cleveland district court lacked jurisdiction because the Greenpeace citizens' suit had come too late in the permit process to be heard. In its decision, the court of appeals did not consider safety, health, or any of the other issues raised in the original trial, ruling only that the permits were valid. Protests continued, but the intensity abated and their frequency decreased. Everything, as we have seen, had happened a little too late for the health of the people and environment in the Ohio River valley surrounding East Liverpool. At the same time Terri Swearingen's monumental contributions to changing the way citizens, industry, and government view toxic waste were acknowledged when she was named by *Ohio Week* as the 1993 Ohioan of the year and by *Time* magazine in 1994 as one of the fifty most promising leaders in America. And more recognition would come, including in 1997 the prestigious Goldman Environmental Prize.

By the end of the 1990s, T-SEC and other protest groups had not been able to shut down the WTI facility—the incinerator was still burning waste. However, the coalition of opposition organizations had played a major role in setting a new agenda for the management of toxic waste. In 1997 for the first time the EPA established siting criteria for hazardous waste facilities and identified eight types of sites to be avoided. By these criteria the WTI site is inappropriate for five of the eight: it is on a floodplain; it has a hundred or more atmospheric inversions each year; it is located over two drinking water aquifers; it is on unstable fluvial sand and gravel; and it is too close to homes and a school.

In the late 1990s the EPA also for the first time set standards for unsafe levels of dioxins released into an environment by hazardous waste disposal facilities. In addition, the agency acknowledged that many toxins move in the food web, making it necessary to assess health risks more broadly than only by measuring the quantity of toxins that a person might take directly from air and water.

Elements like mercury, cadmium, nickel, and lead, and compounds like dioxins, PCBs (polychlorinated biphenyls), and DDT are concentrated as they move in the food web. In the air or water these substances are usually at low concentrations that appear to pose minimal risk—one molecule among a million or a billion. Thus, by dispersing toxins widely, they are diluted to harmless levels. However, plants, algae, bacteria, and other organisms absorb toxins so that toxic elements are more concentrated in organisms than in the environment. Insects and other small animals eat these slightly contaminated organisms and thereby further increase the concentration of toxins. At the top of the food web, species like eagles, polar bears, tuna, and humans are then exposed to concentrations of toxins in their food that are hundreds to thousands of times higher than those found in the air, soil, or water. And these levels do pose serious risks. Studies of salmon—top ocean predators that spend several years at sea—show that we have contaminated the oceans so thoroughly that when salmon return to spawn and die in the headwaters of freshwater streams, their bodies are sufficiently contaminated with a host of noxious substances that headwater streambeds become toxic dump sites. Clearly, most previous health risk assessments are woefully inaccurate.

Initial safety standards set for the release of biologically concentrated toxins appear hundreds to thousands of times too low. For example, according to these standards, water may be safe to drink, but the fish from it might be toxic. Thus, safe release standards for toxins concentrated in organisms can only be established if we know how poisons move in the food web, their levels in our food, and their influence on general ecosystem health.

The constant scrutiny of the WTI facility surely made owners and employees there, and at other incinerators across the country, more mindful of safety standards, maintenance, record keeping, and the overall quality of job performance. And the protesters' broader campaign against incineration in the context of WTI's history must, in large part, account for the fact that no hazardous waste incinerator has been built in the United States since WTI became operational in 1993, nor has WTI added the second incinerator at its East Liverpool site.

In May 1997 the EPA released a 3,800-page document on the four-year, $2 million risk assessment of WTI's incinerator. The report found no significant or unusual risks. Although the incinerator had been operating since 1993, this assessment cleared the way for a final operating permit. Full commercial operations began in June 1997.

WTI opponents had expected this conclusion, but the report did not invalidate or repudiate the troubling data compiled by the EPA and contained in WTI records and other studies. The mercury levels in local children had doubled in the first six months of operation. In the mid to late 1990s two rare types of cancers showed up in local children: a three-year-old girl was diagnosed with rhabdomyosarcoma, an illness that afflicts about one person in twenty million; and two unrelated boys, both under four years old, developed retinoblastoma, a cancer with an expected frequency of 1 in 250,000. The US EPA risk assessment identified 27 different accident scenarios that could result in the loss of life at East Elementary, but harm to residents whose homes were closer than the school was not considered. According to EPA records, between 1993 and 2000 WTI had 27 incidents of unauthorized release of toxic materials, 5 explosions, and 34 fires. Serious violations of EPA regulations included: from 1992 through 1997 air monitoring did not report accurate data because of an improperly programmed computer; between 1994 and 1997 an ambient air-monitoring system had been disabled; and during a lead emission test only lead-free waste was burned. The data indicate that a future fire, explosion, or mishap could cascade into a serious accident for the people in the surrounding homes and school. These data, the authorized and unauthorized emissions, and twenty years of history establish for many people familiar with toxic waste burning that the East Liverpool incinerator is slowly poisoning its neighbors.

Nonetheless, Terri Swearingen and her colleagues had not given up. On Super Bowl Sunday 2000 a fax came into the office of Motel 6 in Tewksbury, Massachusetts, not far from Gore campaign headquarters in New Hampshire. When the clerk saw the cover sheet sporting the United States Presidential Seal along with the words "Executive Office of the President," he smiled and notified the guest of her fax. As Swearingen came toward the counter, the clerk said, "Your friend sure has a great sense of humor!" But it was for real.

Al Gore could still do the right thing. Prior to Superbowl Sunday, Greenpeace and T-SEC had sent him a letter requesting a meeting. When he declined, they sent him and many of their past supporters a letter stating that their members would again resort to acts of civil disobedience on January 31 at Gore's headquarters in New Hampshire. The vice president desperately wanted to avoid responding to acts that would focus attention on the WTI incinerator. The White House fax to Tewksbury's Motel 6 was part of twenty-four hours of nonstop negotiations: Gore had agreed to have EPA's ombudsman review the WTI facility and make a recommendation as to whether it should continue operating, but the recommendation would not be "binding." The opponents to the facility had made more progress in that one day than in the previous seven years. They decided to forgo jail and trust the system once more.

Although the ombudsman's report was to be finished by early summer, it didn't appear until the last week of October. He recommended that the WTI incinerator be shut down for six months while a thorough review was conducted. However, the recommendation had not been binding, and the Clinton-Gore administration did not implement it.

Organized, active opposition to the WTI incinerator in East Liverpool has lasted more than two decades. It has been a knowledge-based, inspired, creative, passionate campaign, yet ultimately it has not achieved what it set out to do. The incinerator was built and has been burning toxic waste for more than a decade, and barring a disaster, it will probably continue to do so for a long time. The daily release into the air of the types of toxins that come out of the WTI stack is unquestionably harmful to both human health and the environment. However, those industrialists, politicians, and ordinary citizens who condone such pollution ignore the message Rachel Carson delivered in *Silent Spring* over forty years ago. Toxins are toxins—release them into the environment where they can move freely, and they will circle back to haunt us in unexpected, devastating ways.

Despite the creative activism and tireless dedication of Terri Swearingen and untold others who rallied with her in protest, the relationship between government and industry puts short-term and narrow economic benefits before human and environmental health and lets all of us down. Fortunately, the failure of the activists in East Liverpool contrasts with the substantial progress in the wider arena made as a result of their efforts to effect change locally. Over the last two decades managing the disposal of toxic waste in particular and garbage in general has undergone considerable changes: incineration is no longer held to be a magical, problem-free process for such disposal. Many communities have learned of the dangers of incineration as a result of the efforts of those opposed to the WTI facility and other incinerator projects. In the capital district of New York state, for instance, citizens under the leadership of Tom Blandy of Concerned Citizens for the Environment, Ken Dufty of Rensselaer County Environmental Management Council, and Judy Enck of New York Public Interest Research Group played a major role in preventing the construction of a huge municipal waste incinerator in the community of Green Island in the early 1990s. Like the WTI facility, this incinerator was to be sited in an economically depressed community and in a river valley with atmospheric inversions. In this instance, citizens mobilized to educate local communities about incineration. Their resistance and careful scrutiny of all aspects of the project slowed government approval. As the project was debated, the economics of incineration appeared unpromising and the incinerator was never built.

Democracy is exactly what Terri Swearingen and other grassroots environmental activists are practicing. They have made a difference, but the challenges are formidable. The WTI case clearly demonstrates that the established institutions of government, industry, and law as well as their applications of scientific information are in need of substantial adjustments if these institutions are to be adequate for addressing the complex environmental situations we face. At the same time, all of us consume or use many things that are produced by means of processes that employ health-threatening materials. And the products themselves are often toxic. The question of what to do with the millions of tons of dangerous chemicals we have produced, and continue to produce, requires an answer.

Many Americans know the stories of hazardous waste released into the environment that has harmed people in the nearby communities of East Liverpool, Ohio; the Love Canal area in New York; Woburn, Massachusetts; McFarland, California; Times Beach, Missouri; and the Hudson River area, to name a few. However, we face a difficult dilemma: will we continue to poison ourselves and all other life forms, or, alternatively, will we strive toward the goal of zero release of poisonous chemicals that can move freely in the environment? Stated this way, the choice seems clear, but then the question becomes, "How?"

The simple answer is to think ecologically about whatever we do. In the nonhuman-mediated biological world there is no "waste;" all discarded "waste" is consumed by other life forms. To solve our waste disposal problem, human beings must imitate this biological world and produce only wastes that can be consumed by other organisms. To demonstrate how this might be done, William McDonough, an architect, and Michael Braungart, a chemist, accepted in the early 1990s the task of making a seat-cover fabric that was also food. At the time chair fabrics were essentially toxic waste; most of the materials in them were mutagens, endocrine disrupters, carcinogens, heavy metals, or persistent toxic substances—feed for nothing but a leach-proof landfill. The group led by McDonough and Braungart screened some 8,000 chemicals used in the fabric industry and 7,962 flunked the food test. From the remaining 38 chemicals the team created a line of chair fabrics that could be composted when worn out—or eaten—with no harmful consequences to the environment or the consuming organisms.

However, scientists, industrialists, economists, and even environmentalists recognize that the human enterprise will always use some materials that cannot feed other organisms. For an example of a creative approach to tackling this reality, consider the half-billion dollar carpet-textile company Interface that relied on fossil fuels for energy and for the bulk of its raw materials, but also used toxic compounds. In 1994 out of concern for the health of both people and the environment Ray Anderson, CEO of Interface, decided a new way of doing business had to be adopted. Led by Anderson, Interface set itself the goal of being a sustainable company by 2020. The CEO sums up this goal—his dream—in the *Interface Sustainability Report* of 1997:

> We look forward to the day when our factories have no smokestacks and no effluents. If successful, we'll spend the rest of our days harvesting yesterday's carpets, recycling old petrochemicals into new materials, and converting sunlight into [the] energy [we use]. There will be zero scrap going into landfills and zero emissions into the ecosystem. Literally, it is a company that will grow by cleaning up the world, not by polluting or degrading it.

In Anderson's vision, Interface would be as close as possible to a self-contained "ecosystem": recycling all of the materials used, not releasing any toxins, and using the sun's energy to power all production. In 2002, compared to the 1996 baseline, the company had reduced the amount of scrap material going into landfills by 79 percent, the amount of water used in its carpet plants by 76 percent, and 8 percent of its energy came from renewable sources. And from 2000 to 2002 the amount of nonrenewable energy to make each yard of carpet had decreased by 20 percent. Anderson believes Interface can meet the 2020 deadline!

Terri Swearingen and her colleagues changed the way many Americans think about toxic waste disposal. They and others emboldened us to envision what was an impossibility until now: a zero emissions world. Others have joined the campaign to establish the possibility of such a world. In our day-to-day encounters—talking with a neighbor, deciding what products to buy, establishing the safety of a chemical, or chaining oneself to a demonstration stack in front of the White House—each one of us can play a role in creating an environment free of intentionally released toxins. We can choose to act in ways that protect or heal all forms of life on Earth, including ourselves, from our current self-imposed toxic assault.

THREE

Restoring Wildlands

Dave Foreman and Preserving Biodiversity

> One of my father's ancestors settled in Calvert County, Maryland, in
> the early 1600s. . . . My mother's family settled in New Hampshire
> before the American Revolution. . . . Folk songs and frontier yarns
> tell my ancestors' tale. I want to believe in their story, I feel my belly
> grow warm with the drinking of their legends, and yet . . . dark
> memories lurk like fading photographs. . . . Blankets given with a
> smile to the Indians of Massachusetts; blankets infested with small-
> pox. . . . Huge piles of Bison bones bleached white by the sun. . . .
> Wagons filled to overflowing with corpses of Passenger Pigeons,
> ducks, and curlews, making their way to urban markets. . . .
> Twenty-five couples dancing on the raw stump of a redwood. . . .
> Five million cubic yards of concrete occluding the flow of the
> Colorado River. . . . Those new beginnings for my ancestors meant
> abrupt, violent endings for countless other forms of life and cul-
> tures. Can we have a second chance at a new beginning?
>
> DAVE FOREMAN, *Confessions of an Eco-Warrior*

In the spring of 1980, five burned-out environmentalists sought what
Edward Abbey, author and wilderness advocate, called "the final test of
desert rathood"—the Pinacate Desert of northern Mexico, in the no-
man's land east of the Baja Peninsula. The Pinacate is a mountainous
desert characterized by its black volcanic soil—searing hot even on
cool days—and by its history of swallowing up lost missionaries and
explorers in the days of the Spanish conquest of the Americas. For
Dave Foreman and the other four wilderness advocates it was the setting for a wild trip of cantinas and beer, coupled with the ultimate
desert experience. They intended to climb Pinacate Peak but ascended
its neighbor, Carnegie Peak. It was only Ron Kezar, a Sierra Club

activist and obsessive peak-bagger, who bothered to climb down and over to Pinacate to leave his name in the hikers' register at the summit. On the same day nine years earlier another entry in the Pinacate Peak register stated, "Clumb up from Tule Tank. Gawd knows why. Neat view. Edward Abbey."

The idea of forming a new environmental group had been brewing before the Pinacate trip as Foreman and his western colleagues watched area after area of wilderness lost to mining, logging, or development. On the way home in Foreman's VW bus, Howie Wolke, Wyoming representative for Friends of the Earth, and Foreman were deep in conversation about their frustrations, ideas flowing nonstop, when Foreman exclaimed, "Earth First!" This summed up what these environmentalists all believed: Mother Earth can afford no more compromises.

A few months later, around a campfire in the mountains of Wyoming, Dave Foreman (then New Mexico representative for the Wilderness Society), Ron Kezar, Bart Koehler (former Wyoming representative for the Wilderness Society), Susan Morgan (education director of the Wilderness Society), and Howie Wolke together formed Earth First!. Its mission was to put the planet's biological health foremost in all decisions, especially those of society at large. Earth First! was not a product of the left, the right, or anarchist groups, but rather a creation of restless youth in mainstream environmental organizations. It was to be different: no dues, no memberships, no formal structure, no tax status, nobody giving or taking orders— just a loose aggregate of people who advocated for Earth and "the great dance of life." It would blend many things, but in the beginning the core participants came from the public lands conservation movement.

Earth First! was founded in response to many frustrations, but primary among them was the failure of established environmental organizations to hold the U.S. Forest Service true to its charge of protecting wilderness areas. In the late 1970s the Forest Service's Roadless Area Review and Evaluation (RARE II) program had not completely listed, nor adequately designated, roadless areas that were to be seriously considered for wilderness designation under the Wilderness Act of 1964. RARE II had set aside in the national forests of the lower forty-eight states 10 million acres as wilderness, with another 11 million acres ear-marked for further consideration. But the bulk, 36 million acres of roadless national forest, was not to be protected by the Wilderness Act and was therefore open for development, mining, and logging. Foreman and others wanted to sue the Forest Service because it had not complied with the National Environmental Policy Act of 1970 (NEPA) in performing its review and evaluation (RARE II), but the moderates prevailed: compromise and work with the political system to get something rather than nothing. These were, and still are, the beliefs and marching orders of mainstream environmental organizations.

Almost a decade earlier, in 1971, Foreman had begun his wilderness advocacy career. Around that time Debbie Sease, another wilderness devotee, gave Forman a copy of Edward Abbey's book *Desert Solitaire*. Soon he was hooked, on Abbey's philosophy and on Debbie. As a volunteer he coordinated the Gila National Forest wilderness campaign for the New Mexico Wilderness Study Committee. At the same time the Forest Service was conducting the first Roadless Area Review and Evaluation (RARE). When Foreman and Sease were not advocating the expansion of wilderness in the Gila National Forest in New Mexico in the halls of government, they hiked its trails to affirm that the Forest Service was doing RARE accurately. In the process they learned the ins and outs of the Forest Service, at the same time becoming aware of how little the agency knew about the lands for which it was responsible. In 1973 the Wilderness Society (founded in 1935 to preserve wilderness in the United States) noticed Foreman's energy and capacity to get people fired up about wild places and offered him a position as its New Mexico field consultant. He jumped at the opportunity and, with Sease, moved to Glenwood, New Mexico, to be closer to the Gila Wilderness, an area designated as wilderness within the Gila National Forest.

Over the next several years the Wilderness Society assembled a powerful force of dedicated, energetic wilderness advocates. Meanwhile the country had begun to grapple with the meaning of the Wilderness Act of 1964 and the Endangered Species Act of 1973. Both were nonpartisan, consensus documents and arguably rank with civil and women's rights acts as among the most noble, humane pieces of legislation passed in modern times. With them, we in the United States broke from our deep cultural belief that humans are the measure of all things. In designated wilderness areas, these acts were meant to ensure that long-established ecological and evolutionary processes are respected and human interference forbidden. Howard Zahniser and other authors of the Wilderness Act intended that the legislation keep these wild places untrammeled and unhobbled—free from human purview. The Endangered Species Act provides that all life has the legal right to exist and that humans, as a moral species dominating Earth's biota, have the ethical obligation not to destroy entirely another species' habitat or the species itself.

Before long, however, the practical meaning of these two acts became clear. Protecting charismatic animals like bald eagles and whooping cranes appealed to a wide majority of the American public, but preserving snail darters, obscure plants, and unappreciated insects was a different matter; defending the existence of little known or unpopular organisms precipitated conflict. Even more problematic became the question of how to preserve the huge expanses of wild habitat needed for top predators—grizzly bears, mountain lions, and wolves. Not only had their habitats been fractured by human development, but also for centuries European settlers and their descendants had advocated exterminating these predators. As we ceased to sanction their slaughter, conflict intensified as these carnivores grew in number and threatened livestock.

Dave Foreman had always been fascinated by big animals. Even before he could read, his mother read him over and over his childhood favorites: *American Wildlife Illustrated* and *Wildlife the World Over, Illustrated,* both published by the American Museum of Natural History. As a child he was enchanted not only by beauty but also by the untamed. He recounts a deeply imprinted experience: "I was seven years old and living in Bermuda. My dad was in the air force stationed there. I was in my swimming suit standing on the beach when a lieutenant in the Royal Navy swam out past the reef and was attacked by a shark that took off both his legs. There was blood everywhere when they brought him on shore. He died two days later in the hospital. That really impressed me. It is a wild world out there. Most people would be horrified by it. I was transfixed. Oh sure, it was scary on some level, but scarier would be a world where there are no wild beasts whose will is theirs, not ours." Foreman sees wild animals as enriching the human spirit, even those for whom we might be prey.

In the mid-1970s in addition to his job with the Wilderness Society, Foreman immersed himself in creative leadership positions with various conservation venues: wilderness chair of the Rio Grande Chapter of the Sierra Club, cofounder and director of the American Rivers Conservation Council, chairman of the New Mexico Wilderness Study Committee, member of the New Mexico Governor's Wilderness Commission, and member of the board of trustees for the New Mexico Chapter of the Nature Conservancy. His seventy-hour work weeks, fueled by his passion and intellectual brilliance, made Foreman highly visible among wilderness preservationists.

As Foreman emerged as a remarkable environmental advocate in New Mexico, Celia Hunter, a former World War II pilot and wilderness entrepreneur, came to Washington, D.C., from Alaska to head up the Wilderness Society, then in financial distress. Hunter brought a sense of adventure and the can-do frontier spirit to the organization. Impressed by Foreman, she invited him to Washington as Director of Wilderness Affairs to consolidate the power of field representatives and to improve their effectiveness by placing them under the supervision of a knowledgeable and inspiring person who would direct their efforts on a national level. Foreman learned not to lobby Congress with mule shit on his cowboy boots, but he never lost his forthright western style and suffer-no-fools disposition. Unable to do more than arrest the financial decline of the Wilderness Society,

Hunter was replaced in late summer 1978. For this organization, the time for the western contingent of dedicated, energetic wilderness advocates in leadership positions had run out. Within six months a frustrated, exhausted Foreman was heading home. Debbie Sease had found the Washington lobbying scene to her liking, however, so Foreman returned to the land he loved without his partner and with a deep belief that the anti-environmental extremists had won. A little over a year later the Pinacate Desert experience inspired Earth First! and Foreman resigned as the Wilderness Society's New Mexico representative.

Foreman's assessment that wilderness and wildlife were taking a beating in the United States is borne out by the events of the early 1980s. In Moab, Utah, on July 4, 1980, not far from Arches National Monument, the setting for Abbey's *Desert Solitaire*, the anti-environmentalist "sagebrush rebellion" illustrated its resolve when a bulldozer decked out with American flags rumbled onto Bureau of Land Management land that had been identified for possible wilderness status. The strategy of the sagebrush rebels—a loose coalition of right-wing extremists, chambers of commerce, developers, and ranchers—was to transfer federal lands to states so that the lands could then be put in private ownership. Ronald Reagan appointed James Watt as Secretary of Interior and Anne Gorsuch as head of the Environmental Protection Agency, giving each the assignment of protecting business interests. The Reagan administration slashed funds for energy conservation and efficiency, and built up incentives to drill the United States out of energy dependency—an impossibility as established by the late Dr. M. King Hubbert who, at the meeting of the American Institute of Petroleum in 1956, accurately predicted that oil extraction in the continental United States would peak about 1970. To obtain this virtually nonexistent oil, wilderness would have to be opened up, and it was. In the twenty-first century this pattern has been repeated as the Bush administration opens up wilderness areas in the lower forty-eight states and Alaska for oil exploration.

Earth First!'s inaugural rendezvous was in Dubois, Wyoming, on the same day in 1980 that the bulldozer rattled into Utah's wilderness. The flyer announcing the Earth First! camp-out stated the group's objectives: "To reinvigorate, enthuse, inspire wilderness activists in the West: to bring passion, humor, joy, and fervency of purpose back into the cause; to forge friendships, cooperation, and alliances throughout the West; to get drunk together, spark a few romances, and howl at the moon." Its second newsletter proclaimed the organization's ultimate goal, or at least Foreman's: create a wilderness preserve system of about 716 million acres that would set aside huge million-acre plus parcels of roadless wilderness for every one of the almost four dozen major ecosystems in the United States. Big places in which natural processes—fire, hurricane, flood, disease, predation, recovery—can occur undisturbed; in essence, places where time-honored ecological and evolutionary processes call the tune without human interference. At this time only sixteen roadless areas of over 1 million contiguous acres remained in the western United States. Most were not protected as wilderness, nor did they contain many of the nation's major ecosystems. To some it may have seemed that Foreman's Earth Firsters had embarked on mission impossible.

In mid March 1981 Foreman walked into Babbitt's hardware store in Flagstaff, Arizona, and purchased three 100-by-20-foot rolls of black polyurethane. The clerk asked Foreman why he needed such huge pieces of plastic. Foreman answered, "We are art students. This is our master's thesis."

"I would like to see it when it's finished," the clerk commented.

"Just look in the paper Monday morning," Foreman told him.

Two days later a huge sheet of plastic was unrolled over a portion of the face of Glen Canyon Dam, creating the appearance of a huge crack in its surface. From a bridge spanning the canyon not far from the dam, one of a cheering crowd of seventy, Ed Abbey bellowed, "Earth First! Free the Colorado!" This rite of spring enacted by Earth First! had brought to life "the monkeywrench gang" of Abbey's novel by the same name.

Following the "cracking" of Glen Canyon Dam, Foreman and Bart Koehler toured college campuses with a road show that preached radical environmentalism. Foreman had an innate ability to hold an audience in the palm of his hand. He encouraged people to "think like a mountain and howl like a wolf," connecting with their wild and animal natures. His message resonated not only with the college crowd but also with others shocked by Reagan-era anti-environmentalism. From 1981 to 1984 Sierra Club membership almost doubled to about 400,000 as environmental groups struggled to hold the line using reason and measured responses. In contrast, Earth First!, with its loosely organized following, was a movement that had immense creative potential beyond more conventional environmentalism. Earth Firsters could demand, "No compromises!" and back up their rhetoric with creative acts of civil disobedience—sitting in trees, chaining themselves to equipment, physically blocking the construction of roads in the wilderness, and in the process through costume, song, and words make fun of their adversaries and draw attention to Earth First!'s causes. The loose organization and informality of Earth First! nurtured individual and small group creativity. This atmosphere lent itself well to monkeywrenching: pulling up survey stakes, disabling machinery, severing power lines, felling billboards, spiking trees. But if accomplished as instructed, none of these things would be done in a way that physically injured people or other forms of life. The group's position of no compromises enlivened and advanced the campaign for conservation biology, providing a counterpoint primarily to radical anti-environmental groups but also to mainstream culture that had, by and large, estranged itself from the natural world.

Foreman and others went to great lengths to separate peaceful civil disobedience and sabotage of machinery from violence against people and life in general. But as Earth First! and its allies began to have some success in slowing development into wilderness areas, the encounters became explosive. In May 1983 Earth First! staged seven blockades to prevent a road from being built over Bald Mountain in the Kalmiopsis wilderness area in Oregon and California. On May 10 a bulldozer tried to back over protesters, almost burying them and stopping only when a woman was covered up to her neck. In response, Foreman and Mike Roselle, one of the founding members of Earth First!, asked the supervisor of the Siskiyou National Forest to cancel the road construction company's contract; however, the supervisor, with compliance of the sheriff's department, maintained that no violence had occurred. Two days later at another blockade a truck driver ran over Foreman and dragged him under the truck, down the road, as Foreman held onto the front bumper. When the truck stopped, its driver, Les Moore, leaned out of the cab window and shouted at Foreman, "You dirty communist bastard! Why don't you go back to Russia?"

From under the truck Foreman replied, "But, Les, I'm a registered Republican." Foreman was arrested for blocking a public road.

These actions focused public attention on the Bald Mountain wilderness controversy, making it possible for the Oregon National Resources Council to raise funds for a suit based upon the failure of the Forest Service to follow NEPA when doing a second evaluation, RARE II. The suit was successful and the road was not built, at least not then.

Foreman challenges people to be passionate, to feel. He declares himself just an animal with the uniquely evolved human capacity to contemplate life and death: "When death comes, I want to enjoy it. . . . Let's not pretend that we're immortal. We're all going to die. . . . If you aren't afraid to die, then you can be happy to live; [if] you aren't afraid to live, you can open up, you can love somebody else and not be afraid of getting hurt. You can love a place and not worry about losing it, because you have the courage to go out and fight for it." His message is to act: take a stand, come hell or high water. If a person acquiesces and runs from commitment, the situation is likely to get a whole lot worse.

Foreman's passion for life and wild things resonated with and inspired many, especially young people. He asked people to believe that they could think like a mountain and howl as if they were wolves. It was, in part, fantasy, but human beings are creatures capable of believing in almost anything—ghosts, space aliens, astrology. Although it is certainly dangerous to deny hard truths like gravity and evolution, misrepresentations of other types of reality may inspire people to accomplish phenomenal things. For example, how many people in the United States, and elsewhere, have been motivated by *Washington Crossing the Delaware,* painted by Emanuel Gottlieb Leutze in 1851, in which Washington stands up in the boat, the stars and stripes flying in the wind? When Washington crossed the Delaware River, the Betsy Ross flag with a circle of stars was months from creation, and on such a crossing passengers do not stand. The time of day depicted in the painting is historically inaccurate, as is the kind of ice. Years after the crossing the picture was painted in Germany as an inspiration to the German struggle for freedom. Should the painting be rejected because it is historically inaccurate? On the contrary, the painting embodies a dedication to a vision and in this way champions noble ideals of freedom and democracy. Foreman's exhortations to think like a mountain and howl like a wolf are calls to the imagination to reconnect us to the biotic world that sustains us.

Foreman loved the unfettered character of Earth First!, but he recognized the organization's limitations, which precluded realization of his larger agenda. By the mid 1980s this radical group had helped the environmental movement to focus on specific issues and had motivated innumerable people to become advocates for wildlands. Foreman realized, however, that a chaotic, anarchistic group like Earth First! lacked the discipline and coherent vision that lasting wilderness preservation and restoration required.

In early 1991 Foreman set out in earnest with Reed Noss and Michael Soulé, both conservation biologists, and several others to formulate an ambitious conservation agenda for all of North America. Doug Tompkins, the owner of Esprit Clothing, funded a get-together of a dozen conservationists in November of that year. The Wildlands Project came to life. The organization would not focus on direct action or sensational protest; rather, it would explicitly articulate Foreman's persistent vision—preservation of the wild—by practicing the conservation biology needed to achieve that vision. The intellectual and scientific roots of the project drew from the nineteenth-century giant John Muir, whose far-reaching conservation ideas and scientific observations of wildlands so greatly influenced twentieth-century biologists, including Aldo Leopold, Eugene Odum, Paul Ehrlich, Peter Raven, Edward O. Wilson, Michael Soulé, and others.

The Wildlands Project is a bold response to the harmful consequences of human activities on other life in general and on North American life in particular. Edward O. Wilson, the evolutionary biologist and an advocate for biodiversity, minces no words in his summary of our past relations with the rest of the biological world.

The somber archaeology of vanished species has taught us the following lessons:

- The noble savage never existed.
- Eden occupied was a slaughterhouse.
- Paradise found is paradise lost.

Humanity has so far played the role of planetary killer, concerned only with its own short-term survival. We have cut much of the heart out of biodiversity. The conservation ethic, whether expressed as taboo, totemism, or science, has generally come too late and too little to save the most vulnerable of life forms.

It is always "too late and too little" for what has been lost, but from what remains springs the passion to make the future different. The Wildlands Project is a heroic venue for radically altering human relations with the Garden of Eden.

The first humans to venture onto the North American continent were the ancestors of the Clovis people, named after Clovis, New Mexico, where Clovis artifacts were first identified in the 1930s. These people were of Asiatic descent, and in their migration east and north over Asia they had acquired the skills and tools to kill and butcher mammoths, woolly rhino, and other large mammals. Near the end of the last ice age, about 13,800 years ago, the land bridge across the Bering Strait opened once again to connect Asia to Beringia (Alaska and north-western Canada). These hunters, who had feasted on big game in Asia, crossed that bridge into a land where mammoth, bison, and reindeer thrived. Several centuries passed before the Cordilleran and Laurentide ice caps to the south and southeast of Beringia melted sufficiently for bands of humans to wander south into what would become Canada and the continental United States. Spread before them was a continent overflowing with life—a grandiose megafauna unaware of the killer emerging from the retreating ice.

American mastodon, Columbian mammoth, long-horned bison, giant sloth, stag-moose, mountain deer, dire wolf, short-faced bear, American lion, saber-toothed cat, scimitar cat, and other giant mammals inhabited the landscape. Male Columbian mammoths could weigh 13 tons, and a species of short-faced bear, at three-quarters of a ton, was probably the largest meat-eating terrestrial mammal ever known. Flesh-eating birds—the largest among them, *Teratornis incredibilis,* with a wingspan of over 15 feet, and New World vultures including condors, turkey buzzards, and black buzzards—soared overhead. Six species of giant tortoise related to those on the Galapagos Islands in the Pacific also lived in North America. The continental fauna comprised numerous small reptiles, birds, and mammals, including a vast number of rodents and at least nine species of cats ranging from jaguars, cheetahs, and cougars to the margay (*Felis weidii*), which weighed just over four pounds. At the time of the arrival of human beings, the Americas teemed with a plethora of beasts, some larger than African elephants and more than a few predators bigger and fiercer than anything now present on our tame Earth.

As the human immigrants from Asia moved south into this virgin fauna, their tool kit of scrapers, cutting tools, and shaft straighteners remained unchanged; then a wholly new invention appeared, the Clovis point. Tim Flannery, an Australian paleontologist, and other scientists believe that this beautiful object and lethal weapon was the undoing of the North American megafauna. With lengths ranging from just under 2 inches to over 9 inches these uniquely designed spearheads were the hallmark of the Clovis culture. In fact, the pioneering Clovis people left nothing else of note. Their culture was omnipresent in North America for the brief 300 years from 13,200 to 12,900 years ago, after which time Clovis points are not found.

During these three centuries this glorious megafauna went extinct except for mammoth, mastodon, and short-faced bears on the ground and California condor (*Cymnogyps californianus*) in the air. These three giant mammals disappeared over the next millennium or so, long before humans even imagined an Endangered Species Act. Condors continued into modern times, and, thanks to a Herculean effort beginning in 1987 that captured the few remaining in the wild and bred them in captivity, a small population of wild condors soar once again over California and the Grand Canyon.

This slaughter of North America's megafauna by humans forever changed the ecosystems in which these animals had flourished, resulting in a great many extinctions. The removal of a predator changes an ecosystem's dynamic balance, sometimes radically. For example, in the nineteenth century the sea otter (*Enhydra lutris*), which lived in giant kelp "forests" off the western coast of North America from the Aleutian Islands in the north to southern California, was hunted for its pelt. Otters feed mainly on sea urchins, so when the otters became locally extinct, the echinoderms thrived and their populations first increased. However, the multitude of urchins decimated the kelp beds until no organisms remained, including the urchins themselves, since they no longer had anything on which to feed. The entire kelp ecosystem collapsed—algae, crustaceans, fish, sea urchins, squid, whales, and other organisms disappeared. For this reason biologists call the sea otter a keystone predator; without it the ecosystem literally falls apart.

The loss of predators can also precipitate changes other than total collapse. Countless examples illustrate this phenomenon, such as the case of Barro Colorado Island in Panama. Before Barro Colorado Island was formed by the creation of Gatun Lake by deliberate flooding during the construction of the Panama Canal, the area supported jaguars and pumas that preyed on numerous small mammals including coatis, agoutis, and pacas. Jaguars and pumas require large territories, and if their habitats are sufficiently reduced, they disappear. Since the new island was too small to support large cats, jaguars and pumas abandoned the island. As a result, populations of coatis, agoutis, and pacas grew tenfold. These seed-eaters became abundant enough to clear the island's forest of large seeds, so trees with small seeds gained a selective advantage and, over the years, claimed a greater portion of the forest. Such changing abundances of species are like ripples on a pond—they create repeating waves of accommodation in the complex web of interactions among species. From top predators to fungi and bacteria, relations adjust to each new set of circumstances. Iterated and replicated throughout Barro Colorado Island's ecosystems, these adjustments changed the ecological landscape of the island, from soil organisms to trees and mammals.

Over time—decades, centuries, millennia—disturbances resolve and new states of quasi-equilibrium are attained. Such was the case 13,000 years ago when North America was cleared of its megafauna. Vegetation patterns changed and surviving mammals migrated. Despite the warming climate a wide diversity of species—including jaguars, llamas, spectacled bears, tapirs, armadillos, peccaries, and ocellated turkeys—retreated south, often to refuges in the jungles and rugged terrain of Central and South America. In Beringia, north of the ice sheets, other large mammals from Asia such as bison, elk, moose, grizzly bear, and wolf headed south into a land lacking large animals but otherwise fecund. Unlike the megafauna they replaced, these northern species had shared their Asian range with human hunters and had evolved habits that made these species more elusive prey. The puma, or mountain lion, appears to have been wiped out in North America, along with the rest of the megafauna, but then surviving populations from the south came to reoccupy the continent.

Over a period of 13,000 years the renewed but impoverished biota and the descendants of the Clovis people and of two later immigrant groups, Na-dene and Inuit, adjusted to each other in myriad patches across the diverse ecosystems of post–Ice Age North America. Extinction continued, but the pace slackened as the most vulnerable were gone and a more equal balance of power between human predator and animal prey yielded local, but not continental, ecological disruption. This uneasy truce abruptly reverted to wholesale slaughter when European civilization expanded westward to satisfy the ever-increasing needs of Europe's growing human population. Armed not only with guns but also with diseases, literacy, and an emerging, scientifically based technology, Western civilization decimated biodiversity and native peoples alike.

Europeans did not adjust easily or instantly to North America, but when they put down roots, they claimed everything around them. Over the past 500 years, the effect of European habitation of the continent has been dominant and lethal. The great auk (*Pinguinus impennis*) in 1844 was the first recorded extinction. Like so many other flightless birds around the world, it was bludgeoned to death for its eggs, meat, feathers, and oil. Martha, the last passenger pigeon (*Ectopistes migratorius*) died in the Cincinnati Zoo in 1914; this species that numbered in the billions in 1800 was gone from the skies by 1900. In a mere 160 years from the first extinction, 20 more bird species would go extinct, and 3 more are missing and possibly extinct in the United States. At the present time, the presumed and possibly extinct species also include: 347 invertebrates, 141 vascular plants, 17 fish, 6 nonvascular plants, 2 amphibians, and 1 mammal. The worst, however, is yet to come. In the late 1990s, a survey of over 20,000 known species in the United States by the Nature Conservancy established that fully one-third are vulnerable to extinction, and human activities are encroaching on the little remaining habitats of other species. In addition to species extinctions, ecological associations are also threatened. These ecological associations are plant communities that represent biological diversity above the species level. For example, two representative ecological associations in Minnesota and neighboring Canada are jack pine (*Pinus banksiana*) with understory balsam fir

(*Abies balsamea*) that grow on moist bedrock substrates and jack pine with understory bearberry (*Arctostaphylos uva-ursi*) that grow on dry sand plains. Of some 4,500 nationally recognized ecological associations established by the Natural Heritage Program, 57 percent are vulnerable to extinction.

Unintentionally, Europeans brought an ecological holocaust to North America, as did the earlier Clovis people. Most of the megafauna that survived the Clovis, Na-dene, and Inuit immigrants went extinct locally with the coming of the Europeans. These creatures hang on in isolated refuges or as cultivated curiosities in parks and zoos. Perhaps a dozen incipient species of grizzly bear were reduced to a few isolated remnant populations in the several large roadless areas in the northwestern and southwestern parts of the continent. Wolves that kept foraging mammals in check across the land have been exterminated everywhere and only survive in remote or protected areas. Mountain lions, bobcats, ocelots, jaguars, and other smaller cats similarly have disappeared from most of their ranges, with few viable populations remaining in isolated pockets of their radically diminished habitats. Two centuries ago, when Lewis and Clark made the first organized exploration across the continent to assess its biological wealth, some 50 million bison thundered across the Great Plains; in 1900 fewer than a thousand survived. Now a tourist attraction in Yellowstone National Park, their range has been reduced to a handful of public and private lands. Across all of North America, human beings are the only large dominant predator remaining. As a consequence of our relentless assaults, the continent's ecosystems are in persistent disequilibrium.

How have human beings wrought such biological havoc throughout the globe? The simple answer is that we have evolved into an organism that is an extraordinarily effective agent of change, our powers practically equivalent to that of photosynthetic bacteria, which, beginning some two billion years ago, filled the atmosphere with oxygen, thereby poisoning most of the planet for those organisms that first evolved in the planet's oxygen-free atmosphere.

Ecological and evolutionary processes established their character in a world where intervals of stasis and extended periods of dynamic equilibrium were interrupted, from time to time, by momentous change. But when humans became behaviorally modern, probably with the acquisition of complex language some 50,000 years ago or more, we gained the capacity to increase the speed of change in the ecosystems we inhabited. Slowly, and in patches at first, we replaced stasis and dynamic equilibrium with change. Before human domination of Earth's biosphere, long-standing ecological and evolutionary processes allowed for renewal in these patches of disturbance, because they were relatively small or few compared to the larger, undisturbed environment. Over the past several centuries, however, the patches of disturbance effected by human beings have become immense; on a global scale we have exerted on the rest of the planet's life forms continuous and ever-changing selection pressures that have made constant, radical change the rule. The speed and degree of such changes make repair of the human-caused disturbances impossible, and severe biological impoverishment has been the outcome.

As human populations increase and ever-more destructive technologies have impact on every ecosystem on the planet, the negative consequences for life in general and for humans in particular are becoming clear. But, to date, any actions taken have been insufficient to arrest the annihilation of many of our fellow creatures and their habitats. At this late hour in North America, only a bold new vision, realized through radical action, appears to have any chance of restoring ecological resilience and arresting further biological impoverishment.

Dave Foreman has such a vision, embodied in the Wildlands Project. The organization proposes nothing less than a reversal of the 13,000-year history of humans as the exterminators of North American life. With the continent's top predators permanently removed and without intervention, the North American ecosystem will either fall apart or further adjust to these removals, which will over time lead to greater impoverishment. The Wildlands Project is based on the premise,

called "rewilding," that returning a full complement of natural predators to the land is necessary to reverse biological hemorrhaging.
Rewilding is simple in concept and grounded in science; the challenge
is in putting it into practice. The Wildlands Project's distinctive contribution is its goal of rewilding a substantial portion of North
America, perhaps half. The project is certainly not the only plan of its
kind, but it offers the possibility of realizing a unique perspective on
how humans might inhabit the earth with less harm to the planet and
to its life forms.

The Wildlands Project has been putting its conservation methods
into practice for almost a decade in the U.S. portion of the Sky
Islands, a biologically rich landscape of forty mountain ranges that rise
above desert and grassland in the southwestern United States and
northern Mexico. The Sky Islands Wildlands Network (SIWN) in
southeastern Arizona and southwestern New Mexico is comprised of
wildlands connected by corridors that will in turn connect to a sister
wildlands network in adjacent states in Mexico. This plan builds upon
the accomplishments of many conservationists, beginning in 1924
with Aldo Leopold's recommendation to the Forest Service that it
establish the Gila Wilderness Area, the first such designation in the
United States. Leopold began his career in the Forest Service in the
Sky Islands region, where the rich biotas of the north and south inter-
mix. Foreman describes the area this way: "[Here] the plants and animals of the Neotropics mingle with those of the Neoartic, where
jaguar and grizzly hunted the same ridges, where elk and javelina
browsed and rooted cheek to jowl, where northern goshawks took
thick-billed parrots on the wing." Many years earlier, after leaving this
region, Leopold described an event on the Blue River that would
inspire generations of conservationists:

> Only the mountain has lived long enough to listen objectively
> to the howl of a wolf. . . . We were eating lunch on a high
> rimrock, at the foot of which a turbulent river elbowed its
> way. We saw what we thought was a doe fording the torrent,

her breast awash in white water. When she climbed the bank toward us and shook out her tail, we realized our error: it was a wolf. . . . In those days we had never heard of passing up a chance to kill a wolf. In a second we were pumping lead into [her], but with much more excitement than accuracy. . . . When our rifles were empty, the old wolf was down. . . . We reached the old wolf in time to watch a fierce green fire dying in her eyes. I realized then, and have known ever since, that there was something new to me in those eyes—something known only to her and to the mountain. I was young then, and full of trigger-itch; I thought that because fewer wolves meant more deer, that no wolves would mean hunters' paradise. But after seeing the green fire die, I sensed that neither the wolf nor the mountain agreed with such a view.

It would take Leopold decades after this Blue River experience, and conservationists in general even longer, to realize that top-predator removal in conjunction with other land management practices was the ruination of the paradise they wished to enhance but could only preserve. The stark proof came to Leopold in the mid 1930s when he visited the Sierra Madre in the Mexican state of Chihuahua. In the region's mountainous areas, the landscape teemed with robust populations of wolves and mountain lions, despite the absence of national parks or nature preserves. In contrast, to the north similar terrains were in ecological chaos. These lands had been severely impoverished by species removal of carnivores, large ungulates, and keystone rodents; by the damming of streams and by riparian forest degradation; by fire suppression; by roads that had fragmented habitats; and by the introduction of alien species. Today the mountain areas of Chihuahua are more like its neighboring states to the north in the United States, with only remnants of its biologically rich past remaining in the Mogollon Highlands, Sky Islands, and Sierra Madre.

Dave Foreman and many others have walked the Sky Islands land to confirm that the map-based design relates to on-the-ground reality.

The conservation aspects of the plan are consistent with the best science and theory available. The SIWN in the United States encompasses over 200 private and public pieces of land that form a mosaic of core wilderness and compatible-use areas connected by corridors. Core areas can harbor complete ecosystems including top predators while compatible-use areas buffer core areas from human impact but allow some economic uses of the land such as very low-density housing, ecologically sustainable forestry, or fishing and hunting. Core areas are connected by corridors that allow migration among core areas. Many conservation initiatives exist in the SIWN that are being integrated into, but not directed by, the Wildlands Project. The role of the SIWN initiative is to provide an overarching model of integration so that networks of people can establish networks of wilderness across the land.

Leopold wrote that the first rule of intelligent tinkering was to save all of the pieces. Conservation biologists are not of one mind on how to save all of the pieces (species, communities of species, and habitats), but considerable scientific evidence establishes that some species— called focal species—can act as surrogates for whole constellations of species, communities, and habitats. In the Wildlands Project, the major focus is to identify, for each given area, the focal species— including native carnivores—and suitable or restorable habitat to support them. The SIWN conservation plan has identified 28 focal species: 9 carnivores, 5 ungulates, 4 raptors, as well as 3 other birds, 4 fish, 2 rodents, and 1 amphibian. If these species can be restored with ecologically viable populations, then their habitats will regain health, as will those of thousands of other species and communities under their ecological umbrellas. When, and if, this restoration happens, a first giant step will have been taken to rewild the Sky Islands.

Meanwhile the Wildlands Project is working with many groups of people on myriad pieces of land to create connecting wilderness areas that run up the continental divide from Mexico to the Arctic Circle, with a loop that includes Baja California, the Coastal Range, and the Sierras, and that connects to mountainous protected areas in northwestern United States and Canada. A similar network is being designed to run from Florida through the Appalachian Mountains to Maine and then to Nova Scotia, with a side branch across the northeastern forest from Maine to the Adirondacks in upstate New York— a 6-million-acre state park—and north into Algonquin Park in Canada. An east-west network across Canada and parts of the northern United States to connect the eastern and western networks is in the early planning stages. When complete, this continental network of core wilderness and compatible-use areas connected by corridors will enable many North American species to live as they once did and to migrate freely, unhindered by human activities.

The Wildlands Project is a master plan complemented by innumerable conservation efforts across the continent. For example, the state of Florida is the first regional government on the continent to aggressively pursue a comprehensive habitat system for preserving its biological wealth, which includes about 100 endangered species, over 160 plant families, more than 700 species of vertebrates, and numerous threatened habitats like longleaf woodland and ravine forests in its panhandle, as well as the Everglades. Satellite photographs in the 1980s revealed to Larry Harris, an ecologist at Florida State University and author of *The Fragmented Forest*, that a huge roadless area existed, running from Okefenokee National Wildlife Refuge in the north along the border between Georgia and Florida to Osceola National Forest in the south, and, between the two, the privately owned roadless Pinhook Swamp. By the mid-1990s over half of Pinhook's 69,000 acres had been purchased by government and conservation groups with private, state, and federal funds to preserve the wildness of a roadless Pinhook. But, more important, saving Pinhook catalyzed major conservation efforts presented in the Florida Game and Freshwater Fish Commission's plan to set aside sufficient habitat to preserve viable populations of 44 focal species. When the plan was written in 1994, existing conservation lands were sufficient to protect just 14 of the 44 species. To provide for the rest, another 4.8 million acres of mostly private lands were needed, which, it was estimated, would cost $5.7

billion. Through the 1990s the state of Florida spent about $3.2 billion acquiring natural areas. With these additional lands acquired as habitats for Florida's many species, the bottom piece of the east coast wildlands network is taking shape.

In 1994 America's next great national park was proposed by the environmental group RESTORE: The North Woods. Maine Woods National Park and Preserve would surround Baxter State Park, and, with 3.2 million acres, it would be the largest national park outside of Alaska. Much of the proposed park consists of huge parcels of hundreds of thousands of acres of logged but undeveloped forest owned by paper companies looking for buyers, so the proposal seems feasible. Detailed economic analysis has established that a national park would provide the area with greater wealth, as well as more economic stability, than the current pattern of use by the forest products industry. In addition, the character and quality of life in rural communities associated with the park could be maintained, if not enhanced, by the thoughtful and persistent involvement of local residents to contain and regulate development.

In communication with RESTORE and numerous other groups in Maine, the Wildlands Project is finalizing its Maine Wildlands Network Conservation Plan. As in Sky Islands it has identified focal species—6 mammals, 3 birds, and 1 fish—and is designing the Maine network of wildlands around the needs of these species. The network would comprise about 48 percent of the state, and about 40 percent of this land would be managed with significant protection for wildlife. The park and Wildlands projects dovetail with the other large-scale planning efforts across the North Woods to connect Maine wilderness, through New Hampshire and Vermont, to Adirondack Park in New York.

When most citizens of North America first hear of returning half of the continent to wilderness, they just give an incredulous look and say, "Impossible." Yes, it does seem highly improbable; yet this visionary goal is worth striving for and believing in, as Edward O. Wilson contends:

Great dreams, as opposed to fantasies, are those that seem to
lie at just beyond the edge of possibility. When I first learned
of the Wildlands Project, I thought it must be beyond that
limit, an admirable whimsy of noble souls. But as quickly as
I gave the idea serious thought, I was converted. With imagination and will, I firmly believe, it can be done. . . . The
return of Nature in swaths of wildlands across the continents
is morally compelling for what it provides future generations.
It is ultimately the best way to protect native faunas and flora,
and to add both physical and biological stability to the global
environment. For the farsighted and courageous, its undertak-
ing will be an epic adventure. For all the rest, its achievement
will increase our security and restore some of the lost prehu-
man magic of the world so vital to the human spirit.

More than two decades have passed since Dave Foreman first proposed saving all of North America's major ecosystems in huge parcels
of roadless wilderness. Time has mellowed Foreman's exuberant, wild
character, but his steadfast attention to this grand task persists. The
belief that all of the major ecosystems in Canada, Mexico, and the
United States could be preserved with connected, million-acre wilderness reserves scattered across the continent is still a dream, but
Foreman may yet participate in the rewilding of many more wildlands
networks. And, if enough people grasp and act upon this vision, we
may begin to reverse 13,000 years of ecological mayhem on the North
American continent and to breathe new life into our dying landscape.

FOUR

Healthy Farms

Wes Jackson and Agriculture

> I think that if we solve the problem *of* agriculture, we can solve
> most of the problems *in* agriculture.
>
> WES JACKSON, *New Roots for Agriculture*

Wes and Dana Jackson cofounded The Land Institute in 1976 as a
nontraditional school dedicated to seeking "sustainable alternatives in
agriculture, energy, waste management, and shelter." In subsequent
years, this mission narrowed to that of replacing industrial agriculture
with a holistic, sustainable agriculture based upon community life:
"When people, land and community are as one, all three members
prosper; when they relate not as members but as competing interests,
all three are exploited. By consulting nature as the source and measure
of that membership, The Land Institute seeks to develop an agriculture that will save soil from being lost or poisoned while promoting a
community life at once prosperous and enduring."

This "nature-as-measure" agriculture is contrary to just about
everything that has happened in farming since Jackson sold fresh produce from his parents' roadside stand in the Kansas of the 1940s.
Although European agriculture may have been ecologically oriented in
the past, the degree to which industrial farming—mechanized, chemically based production of standardized products—has abandoned any
resemblance to a stable ecosystem is phenomenal. Such farming has
become a disturbance-based, human-mediated association of organisms that only persists with massive inputs of artificial fertilizer, pesticides, and herbicides, and huge amounts of energy from fossil fuels.

During the twentieth century, many family farms in the United
States went out of business, and their numbers continue to dwindle.
In 1900 over 50 percent of the people in the United States lived on

farms, whereas today only a small percentage of the population is on
a family farm. Highly diversified farms on a scale compatible with
family management and local economies have been replaced by larger,
one- or two-crop family farms or corporate farms that operate on a
massive scale. The loss of small, diversified farms that serve local markets has many causes, economics foremost among them: efficiencies of
scale and our system of government subsidies economically favor
largeness. All industrial farms incur major expenses for machinery, fertilizers, herbicides, insecticides, fungicides, diesel fuel, and hybrid or
genetically engineered seeds. The expenses incurred by such farms
result in small profit margins, but as farm size increases, so do profit
margins, because increased size allows for economies of scale, especially when the farm specializes in one product such as corn, wheat,
sunflowers, chickens, or pigs. The economy-of-scale savings are, however, dependent upon a heavily subsidized infrastructure of roads, rails,
ocean shipping, and energy that supports long-distance trade at home
and abroad. Small-scale local farmers have little need for this infrastructure, but their products can be undersold because large-scale,
single-product farms can produce a product for less and deliver it
"free" to local markets everywhere.

Modern industrial farming has become a debt-driven business
where short-term returns take precedent over long-term consequences.
Although we may think of the drought years of the 1930s as anom-
alous, erosion rates like those in the first part of the twentieth century
persist. Iowa has lost 50 percent of its fertile topsoil since it was first
plowed in the mid nineteenth century, and it is still losing about an
inch of topsoil every ten years—12 tons per acre per year. In contrast,
undisturbed forests and grasslands have annual erosion rates of .028
and .008 tons per acre, respectively, with annual soil formation rates of
about .40 tons per acre. Current agricultural practices guarantee soil
loss because when soil is tilled (plowed) it is exposed to wind, water,
and splash erosion for substantial periods. Plowing along contours,
using hedgerows to break up fields, planting a cover crop of winter rye
or alfalfa, rotating crops, leaving fields fallow, practicing no-till agri-

culture on sloping fields, and not cultivating marginal, highly erodable lands are all good methods to reduce erosion, but profitably growing annual monocrops—such as corn, wheat, soybeans, sunflowers—on large farms is often not compatible with these practices.

Soil erosion is not the only problem resulting from large-scale modern agriculture. In the United States, the toxic runoff of residues from fertilizers, pesticides, and herbicides is a major source of pollution in lakes, rivers, and ground water. Rachel Carson's book *Silent Spring,* now over forty years old, had a simple message: toxins put in the environment to kill "pests" will remain there and cause unforeseen health and ecological problems. Although she meticulously detailed many specific cases, she described only the tip of the iceberg. The list of documented unintended consequences of chemical use in agriculture is much longer now; yet we continue to create exotic toxins and add them to the environment at rates at least a hundred times greater than in the era when *Silent Spring* was published.

Many Americans believe these deadly chemicals are required to control the numerous organisms that thrive by eating our food. Interestingly, we have, by and large, not reduced crop losses by these measures. In reality, we have to run very fast just to stay in place; that is, most pests evolve resistance to the toxic chemicals that were once lethal to them, while new pests emerge because their natural predators or competitors have been eliminated by insecticides, herbicides, or fungicides. In this way, pesticides often contribute to increases in pest populations rather than reducing or eradicating them. As we have used them, toxins work no better than traditional methods of mechanical removal, crop rotation, and growing a variety of crops in smaller fields with natural vegetation breaks between fields; yet these poisons radically alter and degrade the health of our ecosystems. In fact traditional methods of agriculture, especially those practiced on organic farms of several-to-a-hundred acres that raise a wide variety of animals and plants and functionally mimic natural ecosystems, can give per-acre yields equivalent to chemically based, industrial methods.

Wes Jackson loves the story of the old Sioux Indian who observed an early settler breaking sod. After looking at the furrow with the roots in the air and the stems and leaves under the soil, the Indian said, "Wrong side up." Functionally speaking, the wrong side is up. Jackson is a plant geneticist, not a politician, but he is on a campaign to get the sod right side up. His vision is to create an agricultural system with ecological stability yet no decrease in yield.

Jackson's roots are in the land. He is the youngest of six and grew up on the 40-acre family farm in the Kansas River Valley, west of Topeka. His ancestors arrived not long after the sod had been broken on that farm. As a boy, he found hoeing endless rows of strawberries, corn, potatoes, and other crops in the hot Kansas sun was hard work, but he loved the smell of freshly plowed soil. Newly tilled ground looked so right to him then. When he was fifteen, he went north to work on a cousin's ranch on the prairies of South Dakota along the White River where the sod had never been "busted." Jackson was seduced that summer by an expansive sea of grass and the untamed culture that embraced wildness. He decided that someday he would have that ranch. He never did buy the ranch, but something got in behind his eyes, in his brain, that would be a long time in surfacing.

Shortly after earning his undergraduate degree at Kansas Wesleyan and a master's degree in botany from Kansas University, Jackson enrolled at North Carolina State in Raleigh as a Ph.D. student in genetics. It was an exciting time in that field. Ten years earlier James Watson and Francis Crick had identified the structure of DNA, and in the early 1960s the genetic code was deciphered. Genetics fascinated Jackson. His studies and research went well, and, by the time he graduated in January 1967, universities in Tennessee and Wisconsin were courting him to join their faculties. But despite Jackson's academic achievements, he took a faculty position at Kansas Wesleyan primarily because there he could also coach track, a reflection of his lifetime love of sports.

This was a time of great turmoil in the United States: Martin Luther King, Jr., and John and Robert Kennedy had been assassinated, civil rights workers had been killed by segregationists in the South, and young people were dying in the jungles and on the rivers of Vietnam. Rachel Carson's *Silent Spring,* Paul Ehrlich's *Population Bomb,* and other important books assessed the ecological dangers of the present and foretold a bleak future unless we radically altered our course to embrace a biologically centered worldview.

Although Kansas was not a hotbed of rebellion, students there were testing the system's limits, as they were across the country. Influenced by new ideas and a new social climate, Jackson began to organize what he called a survival studies program in 1968: "I started clipping, tearing, Xeroxing, filing, and seeing all of the 'ain't it awful' stuff—overpopulation, resource depletion, environmental degradation." These three trends came together in what Jackson called "the crisis triangle": as the explosively growing human population passed the 3.5 billion mark with no indication of slowing, our rate of consumption was growing even faster, thereby requiring exploitation of ever-more resources with the resulting pollution of air, water, and soil and destruction of ecosystems everywhere. Out of Jackson's efforts came his book *Man and the Environment: An Environmental Reader,* the first anthology of our environmental crisis.

By this time Wes had married, and he and his wife, Dana, had three children. Royalties from his book enabled him and his wife in 1970 to purchase three acres on the eastern bank of the Smoky Hill River, a few miles from Kansas Wesleyan in Salina, Kansas. Young and full of energy, whenever time permitted, the Jacksons headed out to their bluff above the Smoky Hill, where they began to build a house from scrap and junk.

He had been teaching four years when he applied for a faculty position whose mandate was to organize an environmental studies program at California State University at Sacramento. To Jackson's surprise, Cal State hired him. At Cal State the environmental studies department comprised Jackson and a half-time secretary. Jackson quickly realized that he needed a core faculty for environmental studies. Within six months he had secured and filled two more faculty positions: one full time and one half-time. As enrollment and student interest grew, the program blossomed. In three years Jackson was promoted to full professor with tenure.

Sacramento is in the heart of California's agricultural empire, but for Jackson it was not the soil in which his roots grew. While he and Dana had been in California, they had been adding to their original three acres in Kansas, and by 1973 they had twenty-eight. He requested a two-year leave from Cal State to pursue his dream of homesteading. They made the house overlooking the Smoky Hill River livable, built a barn, and lived from the garden, the chickens, and a cow. The economics were tenuous, financial security nonexistent. In 1975 when Cal State called Jackson and asked, "Look, are you coming back or aren't you?" the Jacksons had a family gathering to discuss things. He recalls, "I said, 'Well, we'd better go back to California.' And my daughter, who in 1975 would have been fourteen, burst into tears, and she said, 'I thought you always said we are not called to success but to obedience to our visions.' So we stayed."

Jackson was almost forty. He still had not found his life's work, but he had decided academia was not the place where he would find his calling. At some level he understood that resolution of the environmental crisis triangle highlighted in his book *Man and the Environment* would not be found through conventional venues. He and Dana talked about starting a school, but their resources were meager: twenty-eight acres, a few handmade buildings, and no money. Despite their few resources, they decided to found The Land Institute and, with financial help from a friend, enrolled seven college students who, for no credit, came to explore "sustainable alternatives in agriculture, energy, waste management, and shelter." On October 17, 1976, five weeks after classes had begun, the building housing the school burned down. Everything went up in smoke: books, tools, supplies, and the building itself. The Jacksons' living room became the classroom, and the applied learning curriculum intensified far beyond what the Jacksons had originally intended.

Throughout the fall and into the spring of the following year, friends, neighbors, students, and family worked to rebuild the class-room. Much had happened over the past year: a wind generator hummed in the almost constant Kansas wind; the financial pinch created by the fire had led to the formation of Friends of The Land, a group that provided financial and other support to the school; and the noted British economist and author of *Small Is Beautiful: Economics as if People Mattered*, E. F. Schumacher, visited The Land Institute. A unique community had come into existence. The fire had refocused Jackson's attention to clarifying the purpose of The Land, the name given to the emerging institute. Jackson's months of rebuilding had allowed his mind to wander across the landscape that he had known for half a lifetime. South Dakota's sea of grass, the unplowed Konza Prairie he had taken students to that fall, the soil-laden waters of the Smoky Hill River that flowed by his window after every heavy rain—all helped him to see the history of degradation of the prairies and a possible response. He realized that industrially based monoculture of annual plants was ecologically unstable and contemplated why and how this pattern of food production had developed.

Beginning some 10,000 years ago in about half a dozen places scattered around the globe, humans independently developed agriculture. One site was in the fertile crescent in the Middle East, where bread wheat originated. Wheat as we know it today, however, was not the original grain. As the last Ice Age waned, two annual grasses, goat grass and einkorn wheat, hybridized to form a grass called emmer. Goat grass and wild einkorn are characteristic of wild grasses in having small seeds that readily detach from the seed head at the top of the plant. Emmer had larger seeds that were big enough to be harvested easily and to provide sufficient nutrition for humans. People harvested emmer seed heads by cutting the stem with a hand-held sickle and gathering the stems with attached heads in bundles to be threshed, but many seeds were lost as they detached from the head and fell on the ground. In time, some of the gathered seeds were saved and "sown" for subsequent harvest. Heads with seeds that tended not to shatter easily as they were harvested were "naturally selected" because the seeds that remained attached were those subsequently sown. Over time, the harvested emmer had more and more seeds that did not break off. As early farmers raised an emmer with a stronger seed head, they bred the emmer with another species of goat grass to produce modern bread wheat, which has even larger seeds on shatter-resistant heads. Human selection pressure had led to wheat seeds that dispersed poorly on their own, so that propagation required humans to intentionally plant the seeds in places where they grew well. Yields were further increased when humans removed from their fields other plants that competed with wheat for nutrients, water, and light. Domestication was complete; human beings controlled wheat reproduction solely for their own benefit.

Intentional planting and breeding of wheat, as well as domestication of other plants and animals, was a momentous fork in the human cultural road. Agriculture gave us the platform from which a host of highly sophisticated civilizations could spring. At its core, agriculture is the intentional control of other life forms to provide for human needs and desires. It profoundly changed human relations with ecosystems: we were no longer constrained by the ebb and flow of materials and processes from the ecosystems in which we participated as gatherers, scavengers, and hunters. As with the domestication of fire, agriculture gave us another almost godlike power over nature. With this power we began to create highly simplified ecosystems from which we could efficiently extract more food and other materials than had previously been available. In ways still not fully understood, more food, combined with cultural changes, led to denser local populations and consequently greater pressures on the land.

Human populations grew slowly at first, until ever-accelerating cultural change over the past million years led to an explosion in the last two centuries from 1 billion people to over 6 billion. For all practical purposes, we have harnessed much of the planet to serve us. Just look down from an airplane almost anywhere and you will see two habitats: city and farm. Yes, you may also see lots of desert; mountainous areas of trees, rocks, and ice; and expanses of forests and grass-lands; but these landscapes are, by and large, of marginal biological productivity as far as human needs are concerned. We have taken for our use, directly or indirectly, about 30 percent of the land's biological productivity, and we have "clear-cut" the oceans: 13 of the 17 major ocean fisheries are in decline, steep decline, or fished out.

The industrial agriculture that currently feeds most of the 6 billion human beings is characterized by herbaceous, annual, monoculture crops grown for their seeds: wheat, rice, corn, sorghum, oats, rye, sunflowers, soybeans, and barley. Each year the ground has to be plowed so that these annuals can take hold and produce seeds in a single growing season. Nitrogen and other fertilizers must be added to maintain soil fertility and thereby crop yield. Weeds need to be controlled with herbicides until the crop has matured sufficiently to shade the ground, and pests that thrive on monocultured crops need to be held at bay by toxic chemicals. Erosion is unavoidable because the soil is exposed to wind and rain for substantial periods of time between tilling and when the crop plants are large enough to hold the soil in place. And soil moisture is lost from exposed soil by evaporation or by rapid runoff after a rain.

Why didn't humans develop a pattern of agriculture to replicate the fertile ecosystems like those of the Great Plains of the United States instead of transforming uncultivated lands into agricultural industrial parks of questionable efficacy? The simple answer is complexity. The prairie is a patchwork of habitats filled with numerous species, each evolved to fill a particular ecological niche and for complex interrelations with innumerable other organisms. In wet years one group of plants grows vigorously, whereas in dry years it does poorly but survives to burst forth when the rains return. Some plants do well at temperatures over 100°F; others tolerate or avoid such temperatures but thrive when temperatures drop. Torrential rains are absorbed by the vegetative cover: thick, intricate mats of roots; rotting parts of leaves, stems, and roots from years past; and by the innumerable invertebrates, fungi, and bacteria that inhabit prairie sod. Soil formation, from geological weathering of rock and the breakdown of organic matter created by the myriad organisms present in prairie ecosystems, usually exceeds erosion. Fruits and seeds, usually small, are produced throughout the seasons. Herbivores, carnivores, and omnivores that forage among this bounty do well, as did native peoples. Simply stated, this phenomenal level of complexity is formidable to replicate by intentional design. Thus, early developers of agriculture did not seek to replicate such complexity, rather they simplified it to produce high enough yields to support their populations.

Modern agriculture has opted also for ecological simplicity maintained by complex technologies that degrade the land. As a consequence, Western industrial agriculture is fragile, and in the long term it is unsustainable. On the other hand, highly evolved, mature ecosystems are resilient and durable over time, except after exceptional events like a large asteroid hitting the planet or major climate change. These data and conclusions were not new to Jackson, but a way to make large-scale agriculture sustainable had eluded him—and everyone else. However, a most unconventional idea began to take hold of him as the emotional intensity of the 1976 fall fire subsided and spring weather awakened the fields of The Land.

As Jackson describes it, "I don't know the source. I do know there was a considerable uneasiness in me that I couldn't explain. And then for some reason a compulsion to sit down and make this table. I can see it yet." Jackson sketched out a chart he had drawn to illustrate his idea. Across the top of the paper Jackson listed four sets of contrasting plant growth patterns: polyculture (many crop plants growing together) versus monoculture (single-plant crop), woody versus herbaceous, annual versus perennial, fruit/seed versus vegetative. Then he added a fifth heading—the current status of those plant growth patterns in agriculture. He started by listing the items in the first row with the last item being current status: "polyculture, woody, perennial, fruit/seed, mixed orchard (both nut and flesh fruit)." In the second row the items in first three columns were the same, but "vegetative, mixed woodlot" were in the last two columns. With four pairs of choices in the first four columns, sixteen combinations were possible (plus the current status of that combination in agriculture), but some combinations were nonsensical: for example, "monoculture, woody, annual, vegetative" was not possible because woody annual plants do not exist. So Jackson struck out the four nonsensical rows. In the twelve that remained, there was one row—polyculture, herbaceous, perennial, fruit/seed—that had no current status in agriculture. In 10,000 years of agriculture humans had not developed a polyculture of herbaceous, perennial plants for the production of fruits and seeds.

Not long after this revelation, Jackson visited Friends of the Earth, the environmental organization that David Brower, after being executive director of the Sierra Club, had formed in the late 1960s. As they were discussing agriculture, Jackson proposed his unconventional idea: most of the problems of till, monoculture agriculture could be resolved if it was possible to create a herbaceous, perennial, polyculture agriculture that would mimic the ecosystems of the Great Plains and that would provide yields of seeds equivalent to those of current monocultures. A handful of young environmentalists were also present at the meeting. It was clear that the young wilderness advocates were not particularly interested in agriculture, which, as far as they could discern, had nothing to do with their mission. Seeing this, Jackson blurted out, "If we don't save agriculture, wilderness is doomed!" The connection was immediately apparent. If more and more wild habitat was put under the plow as industrial agriculture continued to degrade cultivated land, wildlands would be gone. Evidently impressed, Brower asked Jackson to write an article on how to reform agriculture and thereby save wilderness for *Not Man Apart,* a magazine then published by Friends of the Earth.

Wes Jackson's article "The Search for a Sustainable Agriculture" appeared in 1978. Two years later, Jackson wrote in *New Roots for Agriculture,* "I think that if we solve the problem *of* agriculture, we can solve most of the problems *in* agriculture. I think that mixed perennial seed crops can be developed over the next 50 to 100 years, and that they could be sufficiently compelling to be widely adopted." Like building the great medieval cathedrals, creating perennial polycultures will enlist the efforts of generations of dedicated people. Even more daunting are the mind-boggling cultural transformations required to convince people of the rightness of this course; however, Jackson is a patient man. After all, his vision was not formulated in a flash but emerged over half a lifetime.

Walking across a remnant patch of native North American prairie, tall or short grass, reveals just how difficult creating Jackson's vision will be. As you look around your feet, you see plants so tightly packed that little room is available for intruders. Measure a two-foot square. Count the different types of plants within the box, and you will find up to 25 different species. Repeat the process, again listing the plants, and you will find that the list is not exactly the same in the two squares. By repeating this process many times you can calculate that across the prairie hundreds of plant species are growing. The mixture is dominated by perennial plants, mostly grasses but also composites, legumes, and other forbs (herbaceous plants other than grasses). Legumes are especially critical to the mix, because they harbor bacterial symbionts that fix atmospheric nitrogen to provide usable nitrogen for the legume and, ultimately, for the rest of the prairie's organisms.

Climate, macro and micro, determines the particular species present in any patch of prairie. In the Canadian prairies, cool-season grasses dominate, whereas farther south, in Kansas, warm-season grasses prevail. The tall-grass prairie that once constituted the eastern third of the Great Plains has three dominant grasses—bluestem (*Andropogon gerardii*), Indiangrass (*Sorghastrum nutans*), and switchgrass (*Panicum virgatum*). In the cooler Dakotas, needlegrass (*Stipa* species) and wheatgrass (*Agropyron* species) are also prevalent. Farther west lies a more drought-tolerant short-grass prairie where grama grasses (*Bouteloua* species) and buffalo grass (*Buchloë dactyloides*) are common. In the drought of the 1930s, western wheatgrass (*Agropyron smithii*) and other short grasses moved east to replace tall grasses, only to retreat westward when the rains returned. This kaleidoscope of species distribution seen over time and space characterizes, and is the basis of, prairie resilience and permanence.

Prairie grasses have longish, thin leaves that reach up from the soil toward the sky and flower spikes that crown elongated stems. Shoots are more often clumped than single. Grass roots form a tangled mass that spreads out extensively in the upper feet of soil, but some roots penetrate a yard or two down. Leaves of legumes, composites (such as sunflowers, daisies, coneflowers), and other forbs are generally broader than those of grasses and can lie flat on the surface or appear in various patterns on an elongated stem. Stems may be single or branched with a diversity of floral inflorescences displaying single or multiple flowers. The roots of forbs may remain in the upper soil, but often a taproot, usually as thick as or thicker than the stem, penetrates deep into the soil, sometimes a yard or more. Uncountable invertebrate, fungal, and bacterial species are found above and below the ground in loose or intimate association with this flora.

These plant architectures and associations optimize not only the acquisition of sunlight and retention of rain water, but also extremely tight and effective nutrient cycles whereby calcium, carbon, iron, magnesium, nitrogen, potassium, phosphorous, sodium, sulfur, and other minerals are passed from one organism to the next in the food web. This complexity results in energy flowing through the prairie in a highly efficient manner that supports innumerable organisms and the movement of nutrients from one organism to another, with minimal loss, into the abiotic environment. The prairie is essentially a self-sustaining system that runs on sunshine and uses carbon dioxide, nitrogen, and water, as well as the nutrients in the soil, which are gifts of geological activity: volcanic eruptions, glaciation, and more recently the weathering of uplifted soil and rock.

Since most prairie plants are perennials, considerable resources are put into their roots, the long-lived part of the plant. If one compares the roots and the above-ground growth in a patch of prairie, 60 to 75 percent of the biomass will be roots. These species reproduce by budding from roots and underground stems as well as by seeds, and as a general rule, perennials produce small seeds. These patterns of resource allocation in most perennial plants are not desirable for human food production since we cannot digest cellulose, the compound that makes up the bulk of grass roots and shoots. In addition, the seeds provide too little food by modern standards to be economically harvested.

The Great Plains fed native peoples well, not only because they were master gatherers of edible plants, but also because they hunted the huge herds of bison and antelope, as well as other smaller animals, that converted prairie plants into nutritionally rich meat. Of course, there are many more people now living off the Great Plains than there were native peoples then. However, despite commonly held beliefs, industrial agriculture does not increase human carrying capacity (the number of people that can be supported by a place in perpetuity). Soil erosion; the draining of aquifers; and fossil fuel for tractors, fertilizers, herbicides, insecticides, fungicides, pumping water, seed production, and transportation—as well as the considerable requisite infrastructure—all make the yields we now harvest from the Great Plains unsustainable. History and the natural sciences tell us that we are reducing at an ever-accelerating rate the stocks of natural resources—fish, forests, soil fertility, ground water, and so on—that provide for tomorrow's carrying-capacity. Conservative analyses indicate that, on a global scale, in 2000 we were using all the productive capacity of Earth's ecosystems and the equivalent of about 30 percent more by reducing stocks of natural resources, thereby diminishing future productivity. With the population projected to increase by 50 percent and consumption by 100 percent by 2050, natural resources that provide life support are almost certain to decrease substantially. Simply put, we are accumulating ecological debt by using up Earth's future carrying capacity, which will be paid by future generations.

The challenge taken on by The Land Institute in 1978—the creation of an ecologically based prairie agriculture—is a megaproject in intentional ecosystem creation. The complexity of this task can be appreciated by listing exactly what has to be accomplished:

1. Identify plant species representative of the dominant species found in a native prairie, each with a potential for substantial seed production.

2. Breed these species individually to produce large seeds so that when the plants are grown as a polyculture the overall harvest will be similar to that of prevailing monocultures.

3. Breed for flowering times and seed head morphologies that permit mechanical harvesting and for seeds that are millable into flours with appropriate textures for cooking and baking that also taste good.

4. Grow groups of plants bred for these traits as an ecosystem to establish that the reconstituted ecosystem will maintain integrity under the range of environmental conditions experienced in the areas where these groups will be grown.

5. The assembly of plants must resist invertebrate and microbial pests, require no nitrogen fertilizer, and when planted together, self-assemble into a stable ecosystem that maintains high seed yields over many years while efficiently flowing energy and cycling nutrients. And during all of these manipulations the individual species must remain perennial.

Although this type of intentional ecosystem creation seems novel, it has a precedent. In the 1970s, systems ecologist John Todd proposed assembling ecosystems into what he calls "living machines" to clean up various human-produced wastes, from standard sewage to industrial effluent. After little more than a decade's work, Todd demonstrated that he could bring a host of organisms together and have them establish complex, functional ecosystems to convert effluent to harmless products; human sewage, used malt from brewing beer, or excess fats from ice cream production could be biologically converted into oxygen, water, carbon dioxide, and other benign compounds. Making a prairie agriculture will, however, be much harder. The goal here is to have a specific group of plants that self-assemble into a functional prairie in which these species survive and produce economically harvestable quantities of seeds over many years. Although The Land has been working on this project for almost three decades, from the outset Jackson guessed that it would take the better part of a century to achieve this goal.

In spring 1978, Jackson's group selected 35 varieties of perennial grasses and forbs as possible candidates for seed production and started growing them. Because The Land was a school, its primary function was education, so the Jacksons and their students—there was no staff—became a community of learners. To bring people, ideas, and celebration to The Land, the Jacksons organized the first annual Prairie Festival—"Prairie Roots, Human Roots: The Ground of Our Culture and Agriculture"—in late spring 1979, and around harvest time that same year, the first annual Fall Visitors' Day drew a crowd of over eighty. During the first years of operation, some of the most important environmentalists of the mid twentieth century—including Wendell Berry, David Brower, Amory Lovins, Frank Popper, Paul Sears, and John Todd—shared their visionary agendas in The Land's applied educational curriculum.

By 1982 the Jacksons were overwhelmed with the tasks before them. Breeding 35 wild perennial grasses and forbs had yielded thousands of seeds to plant and evaluate. At the same time they needed baseline data on native prairie to compare with monoculture productivity, as well as data on how the selected perennials grew individually and in various patterns of association. Eastern gama grass, wild rye, and Illinois bundleflower were chosen as the three herbaceous perennials for extensive study. Patches of native prairie came under the microscope for multiyear studies to assess plant composition, biomass, and other characteristics, seasonally and over decades. The Jacksons found that there was much work and few resources of people, time, and money. Over the next few years, research grants and gifts allowed them to hire several staff members to complement volunteers and to give continuity to student projects. Prairie Festivals, Fall Visitors' Days, symposia, visiting scholars, a fresh batch of interns each year, new staff, and visitors of every stripe—all perpetually invigorated The Land's community of learners. The Jacksons had started a school not as gurus for the ignorant but to honestly explore fundamental issues and questions for which no one had complete answers.

Jackson had long believed that "any natural ecosystem was sure to improve, and by that [he meant] add top soil, increase in stability, maybe diversity, or if not improve, at least stay good indefinitely." And he believed this until he made a field trip in September 1985 near Comptche in Mendocino County, California, where the ecologist Hans Jenny walked him up and down Mendocino's ecological staircase that begins near the coast with a vibrant redwood-fir forest and cul- minates inland at higher elevation with a stunted, impoverished pygmy forest. Over time, mountain building had sequentially raised each "stair" from the sea, elevating the pygmy forest now at the top, which is half a million years older than the redwood-fir forest at the bottom. With four terraces the staircase traces a half-million-year suc- cession of soil and ecological impoverishment.

Jackson recalls this experience with humility: "Fundamentalism of any variety tends to die hard. Staring into a soil pit dug into the fourth terrace [of the pygmy forest], I could sympathize with the churchmen who refused to look through Galileo's telescope. Even there, with the evidence before me, I protested, saying that good farming can improve the soil." The hard truth is that over time soil becomes impoverished— in ecosystems unperturbed by humans and even in agricultural set- tings that have been managed to minimize impoverishment. Jackson's quest goes on, but the lesson given that day made him even more skeptical of finding in nature absolutes that could be easily incorpo- rated in a durable agricultural enterprise.

The more we learn about life, from the molecular structure of the gene to the functioning of the biosphere, the more complexity we uncover.

Even what we know now is dazzling and almost unbelievable; no one human mind can grasp it all. In the last few decades, however, researchers have been keen to understand this complexity and to discover why order appears in multicomponent, self-regulating systems like ecosystems.

The Land Institute held a four-day symposium in January 1994 titled "Complexity in Ecology, Agriculture and Medicine." James Drake and Stuart Pimm, then both at the University of Tennessee, provided invaluable insights on ecosystem assembly. Their modeling analyses demonstrated two principles: the order in which species enter an ecosystem determines the outcome at a particular time, and species that go extinct during the assembly of an ecosystem are critical to the species composition at any given moment. For those that have gone extinct, their initial presence has modified habitat characteristics so as to influence the success and failure of other species. These insights indicated that The Land staff could catalogue and measure everything in the native prairie plots, but not knowing what species had disappeared in the past, staff members would still be in the dark about how the system came to work the way it does. Clearly, having many of the pieces of a prairie does not enable one to intentionally remake the prairie. A prairie agriculture appears to be an assembly that is the result of an interactive process of species coming and going over a long period of time that settles eventually into a dynamic equilibrium that may, or may not, persist depending upon the vagaries of environmental change. To remake a prairie one must determine how many plant species are needed and in what order they must come together to produce the intended result. The difficulty in designing such an agriculture lies in the attempt to replicate complex and myriad natural processes that have occurred over a long period of time.

By 1994, Land Institute researchers were working with four species— eastern gama grass, a drought-tolerant grass; wild rye, a drought-sensitive grass; Illinois bundleflower, a legume; and Maximillian sunflower, a composite—but clearly there were too few. Stuart Pimm summarized: "there is not better than a snowball's chance in hell of getting those species to work as four species, simply because the probability of finding a complex system that works is small. The way to get a complex system to work is to do what we call a shake-down, . . . actually get a lot more species in the system."

That spring the polyculture shake-down experiments had groups of 4, 8, 12, and 16 species, with each group adding one more representative from each of the four basic types. Data piled up, some of it encouraging. After three years, plots planted with 12 and 16 species had become dominated by those perennials, whereas those initiated with only 4 perennial species contained 75 percent "weedy" annuals. Other data were discouraging. In two-year-old polycultures of the 4 workhorse species, Maximillian sunflower apparently inhibited the yields of the other 3 species, so that the overall yield for the polyculture plots dropped below that of monocultures of those 4 species, in conflict with earlier results. Something different in the conditions of the plots at the start of the experiments or unknown environmental variations between the plots over the two years had led to disparate outcomes. As the researchers had anticipated, complex adaptive systems do not yield their secrets readily.

Although the theoretical concept of a prairie agriculture that yields like a monoculture yet causes no soil erosion and eliminates the need for nitrogen fertilizer and biocides is appealing to many farmers, it remains an abstraction. Reality for the farmer is planting this year's crops. At the same time, The Land's holistic agenda lets farmers get used to the concept and participate in the creation of polyculture plots. In spring 1994, 77 farms located in 14 states from the Rocky Mountains to Lake Michigan initiated The Land's Great Plains Research Project by planting polyculture plots on their land. Although it's unlikely that anything of commercial value will come from these plots, they will provide unique and relevant data. More important, their existence serves to disseminate among farmers the idea of a prairie agriculture.

Talk, however, gets stale when action is too long delayed. Jackson knows this; therefore, over the past several years he has initiated a new effort to perennialize several major crops that are annuals—wheat, corn, sorghum, soybean, and sunflower. All of these crop plants have close wild relatives that are perennial, and some, like wheat, have over the past century intermittently been bred to create perennial varieties of commercial value. Wheat scientists at Washington State University have joined Jackson's effort, and perennial wheat now grows in The Land's greenhouses. Although the challenges to growing commercial perennial wheat in the United States are substantial, the agronomic tools to do so are well established. With the creation of a commercially viable perennial of any major crop, the conversations among farmers will change from "What if . . . ?" to "How much?"

Although Wes Jackson was credited with coining the term "sustainable agriculture" in his *Not Man Apart* article published in 1978, he believes the sustainable agriculture agenda as it stands is transitional. He calls his program "Natural Systems Agriculture." The title of the 18th Prairie Festival, "The Marriage of Ecology and Agriculture, the Next 20 Years," highlights the difference between sustainable and natural systems agricultures. Jackson believes the sustainable agriculture community's work of reducing the use of chemicals and other impoverishing aspects of industrial agriculture is necessary and important but, on the whole, falls short of aggressively addressing the fundamental problem *of* agriculture itself. A durable, or natural systems, agriculture will have evolutionary and ecological principles at its core. To the extent that it follows the natural principles of biology, it will endure; to the extent that it attempts to get around those principles, it will fail.

Agricultural appropriations legislation signed by President Bill Clinton in October 1995 states that the secretary of agriculture is to "make an analysis of the feasibility, productive potential, and economic and environmental benefits of long-term natural systems agriculture and to identify associated near-term research needs." Although the ideas embodied in The Land's Natural Systems Agriculture program gained recognition in high places, Jackson realized that the program would make few inroads into the high-stakes game of industrial agriculture unless it played on the same field. With this in mind, he recruited over a hundred accomplished scientists, academics, and practitioners to form the Natural Systems Agriculture Advisory Board to spread the word and encourage relevant research. Next, The Land raised funds to support graduate-student research fellows to pursue their research at universities while working closely with The Land. Nearly two dozen fellowships have been awarded, and fellows conduct natural systems agriculture research in universities throughout North America and in Australia. Now Jackson is planning a major natural systems agriculture research center at The Land, with a 25-year plan to make ecologically based farming the standard.

At the 1997 annual meeting of the Ecological Society of America, The Land's Natural Systems Agriculture group organized a symposium on natural systems agriculture that was highlighted in an article by Stuart Pimm in the international journal *Nature*. As The Land creates an ever-expanding network of colleagues and advocates and as natural systems agriculture research papers are published more frequently than a decade ago, the program's momentum is building.

Wes Jackson has known from the beginning that even if The Land's staff mastered the science that would enable them to create a prairie agriculture, the knowledge would be useless without addressing the associated changes needed on the farm and in society. Most farms today are supported by an extractive economy that uses fossil-fuel energy and other energy-dependent resources mined elsewhere to run their operations. Is it possible to run a farm on sunshine that is equivalent in output and efficiency to the extractive-economy-based farm?

In 1993 Jackson started the 210-acre Sunshine Farm as a ten-year experiment to assess how and to what extent a cereal-grain, livestock farm can be run on sunshine. Heavy-duty field operations like plowing are accomplished by a diesel-engine tractor run on vegetable oil (biodiesel fuel), while draft horses do the light-duty field work. In the second year, grazing experiments began when eight long-horn cows, whose breeding history makes them ideally suited for living on native plants, were released onto the 160-acre, native-prairie part of the farm. The cattle, like all animals in an ecosystem, are critical for recycling nutrients, and in this case they also provide a protein-rich food. The herd was allowed to grow until about fifty cows, yearlings, calves, and a bull grazed the prairie. Two decades of experimental work on perennial polycultures inform the planting, harvesting, and rotation of perennials at the farm. Marty Bender, the lead Sunshine Farm scientist, has published research papers on biodiesel fuel; the economics of traditional Amish agriculture; plant uptake of nitrogen, phosphorus, and potassium from the soil; the potential for conversion of biomass to plastics, solvents, and other synthetic organics; and other topics. With the ten-year study completed in 2002, results indicate that about 85 percent of the annual energy required to run a farm can be provided by on-farm, solar-energy flow initially acquired by cultivated plants that are used to feed the animals, make biodiesel fuel for the tractor and off-farm transportation, and fertilize the soil and other energy requiring processes. When the energy to make the tractor and other equipment used on the farm is part of the calculation, about 50 percent of the energy required by the farm can be provided by the farm itself.

Sunshine Farm was created to keep the ecological accounts of energy and material flows in agriculture so as to establish the feasibility of an ecologically based agriculture, but every farm is part of a community. Farms are in fact the foundation of many rural communities, and their economic well-being often depends on these farms. Industrial agriculture, however, promotes the deterioration of rural communities. Since the densely populated cities and urban areas where most of us live depend upon rural communities and towns to provide our food, it is imperative that farm communities in decline be revitalized. This renewal will return people to the land and enhance the stability of our food system. Jackson sums it up this way: "We have to increase the eyes to acre ratio." Rural agricultural areas need to be repopulated with a diversity of people whose skills and talents will make these areas desirable places to live. Jackson calls the repopulation "homecoming" and has proposed that universities establish a new degree program in homecoming to educate young people about returning to the land. This proposition and the challenges involved in creating such rural communities have permeated The Land's agenda from its inception.

Matfield Green—a small, slowly dying rural town in the Flint Hills of Kansas—has become a testing ground for homecoming. Located in Chase County, the geographical center of the lower forty-eight states, Matfield Green is being used to set up the ecological books to see where everything comes from and where it all goes. This is an experiment in a type of accounting called ecological footprinting. The concept is simple in theory but extremely complex in application. Ecological footprinting enables us to assess crudely how much ecological space is required for a given lifestyle to be sustainable. To be more specific, the ecological footprint of a population is the total area of the terrestrial and aquatic ecosystems required to provide the resources consumed and to assimilate the wastes generated by a population in a particular location at a specific time. Take a place like a university, a town, a watershed, or a country, and put a bubble over it. If the place, over time, can be entirely self-supporting under the bubble, it can be described as "sustainable," because in providing for its population it does not degrade the environment. Conversely, a community will collapse over time if it is eroding soil fertility; changing climate; polluting air, soil, or water; reducing life-support capacity by ecosystem, habitat, or species loss; or causing human disease. A community where any of these conditions exist cannot be called "sustainable."

However, because it is physically impossible to put an impermeable bubble over a unit, calculating a footprint becomes a task of listing everything that occurs internally within a community, as well as everything that moves in and out. This was done for the city of Toronto, Canada, in 2000, with its population of 2,385,000 occupying an area of 245 square miles. It was calculated that each person in Toronto requires an estimated 19 acres of ecological resources and assimilative capacity. This translates into an ecological footprint for Toronto of

71,735 square miles, or an area about 290 times as large as the city. This means that a huge number of people can live in a small area because they have access to ecological space everywhere else on the planet— their food, fuel, and most other resources come from lands outside the city limits. A conservative assessment in 2001 established that humans would require at least three additional Earths if all 6 billion people on the planet lived as North Americans do and wished not to compromise the life-support capacity of future generations.

The project in Matfield Green goes beyond ecological footprinting to try to understand how rural communities work and what makes them desirable places to live. Wes Jackson deeply believes that people will come home to the land if place-based communities escape from the bondage of industrial agriculture and its extractive economy. Just how this will be accomplished is unclear. The Matfield Green project is Jackson's venue for exploring possibilities for revitalizing agricultural communities.

I first met Wes Jackson in the fall of 1992 when he lived in a Matfield Green house he had purchased and fixed up. At the time, The Land, Jackson, and like-minded people were restoring the town. For a while the old hardware store served as the Matfield Green Cafe where folks came and chatted about the way things were and the way things are today. The Land's offices then moved into the hardware store, and the refurbished grade school became a town meeting place that hosted all kinds of community functions including regular pick-up basketball games among local kids.

The Land's first major outreach program in Matfield Green was the Rural Community Studies program comprising three central Kansas school districts. The names of the program's workshops and symposia tell the story: "Prairies, Plants, and People of the Flint Hills" connected history and ecology; "The Tall-grass Prairie: Reading the Landscape of Home" revitalized the relationship of people to the land; and "The Significance of Watersheds in Shaping the History, Geology, Archaeology, and Natural History of the Prairies" reminded people that they and their land are a product of natural processes. Dozens of school projects—transforming a city vacant lot into a prairie park, planting 500 trees as a windbreak near the school, restoring 5 acres of European grasses to tall-grass prairie that will become the school's "classroom" for research and prairie management, researching and creating a video of the history of the town's courthouse, landscaping the school grounds with native plants, keeping year-long journals on the flora and fauna in the school's outdoor environmental center, researching the town's history and with local artistic talent creating a town history mural in the school—all connect students, teachers, and others to place.

I attended my first Prairie Festival in 1998. On an excursion around The Land Institute with Jackson, we drove by an alfalfa field heavily infested with weevils. He pulled into the shade of a grove of trees next to the field, turned off the engine, and said, "Alfalfa looks pretty bad. We allowed a local farmer to use this field. Part of the deal was no insecticides. Now that he's got a bad weevil problem, he wants to spray. Look, that's the problem. Say, we let him spray. It seeps into the ground water, travels to our neighbor's well, and nine years from now she has breast cancer." Jackson drew a circle in the thick dust on the dashboard and continued: "This is the boundary of consideration." He then drew a second circle around the first: "This is the boundary of consequences. As it stands now, that breast cancer is here"—he pointed to the space between the two circles—"outside the boundary of consideration. So, where do you draw the boundaries? We told him he couldn't spray." Jackson started the engine and drove on while we discussed boundaries.

Jackson wants people to think hard about the boundary of consequences that we have accepted with industrial agriculture in general and with till agriculture in particular. The Land has taken on the daunting task of expanding boundaries of consideration in agriculture so that they are expanded beyond mere efficiency and short-term profit to include and respect ecological and evolutionary principles.

After all, the larger sphere is biological, not economic. Unfortunately, our economics-based culture will continue to find it exceedingly difficult to set boundaries that accommodate such principles, primarily because Western culture has contrived an economic system that does not respect or consider the laws of nature. Taking a somewhat longer sweep of history, Jackson sums up this way, "[W]e of Western civilization have moved from the church, to the nation-state, to economics as the primary organizing structure for our lives. We have been through the hypocrisy of the church, the atrocity of the nation-state that peaked with Hitler, and now we are devotees of economics, the encoded language of human behavior that directs us toward ecological bankruptcy. It is time to move more aggressively on to the fourth phase, already under way, ecology."

What makes Wes Jackson a visionary is that he is not distracted by short-term, nearly manageable issues. Rather, through endless action he is obedient to his distant objectives. Jackson's work, as embodied in The Land Institute, is the harbinger of a twenty-first century agriculture that could not only save the remnants of good soil and wilderness, but also return much of our impoverished agricultural land to health by adopting ecological and evolutionary principles.

FIVE

Living Locally

Helena Norberg-Hodge and Globalization

Western society today is moving in two distinct and opposing directions. On the one hand, mainstream culture led by government and industry moves relentlessly toward continued economic growth and technological development, straining the limits of nature and all but ignoring fundamental human needs. On the other hand, a counter-current, comprising a wide range of groups and ideas, has kept alive the ancient understanding that all life is inextricably connected.

HELENA NORBERG-HODGE,
Ancient Futures: Learning from Ladakh

Most of us live among thousands of strangers, our physical needs met by the activities of countless other unknown people in distant places. In Western society we believe global integration is inevitable and good. In the most recent period of human history this arrangement, however, has happened so fast that it can hardly be considered a natural pattern. In fact, when we look back at human origins and study traditional cultures and languages, we see that our current relations with each other and with the rest of the natural world are at odds with much of our genetic and cultural heritage. We evolved as a cooperative small-group animal; and each group employed a distinct language and participated in a unique culture that associated the group with a particular place and environment.

Our genetic heritage, like that of all extant organisms, traces back some 3.8 billion years to the origin of life on Earth. Human uniqueness began in earnest some 6 million years ago when the *Homo sapiens* line split off from the lineage of our closest living relatives, bonobos (*Pan paniscus*) and chimpanzees (*Pan troglodytes*). In another 3 million years the bonobo and chimpanzee lines split. These recent separations mean not only that all three species share many genes—at least 99 percent are held in common—but also that we share many behaviors. As evolutionary theory predicts, we see "humanness" in our close primate relatives and, alternatively, "bonoboness" and "chimpanzeeness" in human behavior. In fact, as we learn more about the behavior not only of our closest primate relatives but also of other social animals, we realize that very little of our behavior is not observed in other animals. So, other than the degree to which toolmaking, cooperation, aggression, and other behaviors are expressed, is there anything fundamentally different from other animals in our behavior that has led to our spectacular success as a species?

Although not all will agree, I am persuaded by the argument of some who suggest that language in general, and syntax in particular, has been the wellspring from which we have acquired our distinctive character. Even single-celled bacteria can communicate with other life and the environment by sending and receiving signals that elicit particular responses. But human language is different. Spoken words are abstract symbols for the real and the imagined. Strings of words, with varied order and pronunciation, enable us to represent and analyze the real and the imagined with nuance and precision. Language has given us the capacity to assess the past and to project into the future. We create fantasy worlds just for fun or to give explanation to what we don't comprehend, and we construct relationships, real and fanciful. Like no animal before us, with language we gained the capacity to acquire the sophisticated tools that would enable us to understand, albeit incompletely, the biological and physical principles that created us and the biotic enterprise of which we are a part. Spoken and then written language may have been the fundamental tool that enabled us, a rather physically impoverished primate, to become a pervasive ecological and evolutionary force on the planet today.

Traditionally, a specific group of humans tended to live in a particular location. An indigenous people's language is intimately associated with place and connects the people not only to their particular environment but also to their culture. In addition, their language and how they use it reveal their worldview—how they understand the world about them, their place in it, and their interactions with it. Of course, languages and cultures do change over long sweeps of time. Over the past 50,000 years humans have spoken thousands of languages, but most have vanished along with the people who spoke them. And with the disappearance of a people and their language have gone records of how they lived and related to their places.

In the past these losses have been gradual. Recently, the rate at which languages and cultures are disappearing is unprecedented. Western languages—primarily English—and industrial culture are rapidly homogenizing place-based societies all over the globe. The profound long-term consequences of these changes are not yet known, yet it is clear that with the loss of indigenous languages comes the loss of place-based knowledge. As a world culture we are discarding our appropriate relations with the natural environment of which we are a part; associations that have nurtured and are successful over millennia are being abandoned.

Although these changes are pervasive, they have been difficult to grasp, in large part because neither those people who are changing nor those bringing about the change are able to readily acquire a holistic understanding of the process or its value. Although rare, a holistic view can emerge when the same person gains the perspectives of both those undergoing change and those who are agents of change. Helena Norberg-Hodge is such a person.

Helena Norberg-Hodge grew up in Sweden. Relatives in Germany and England—her mother was half-German and her father half-English—gave her a broad cultural experience and exposure to a number of languages as a child. She studied Spanish in Mexico when she was nineteen, and, by the time she graduated from university, she was fluent in English, German, Italian, Spanish, and Swedish. In the summer of 1975, while working as a linguist in Paris, Norberg-Hodge signed on for a six-week job with a film crew. The project was a documentary film on Ladakh, or Little Tibet as it is often called, located in the northern Indian states of Jammu and Kashmir on the Pakistan border. At the time Ladakhi was a spoken but not a written language, and few, if any, westerners had learned Ladakhi.

Norberg-Hodge soon became enchanted by the place and its people. Ladakhis exhibited a tolerant, contented demeanor that elicited mutual respect from the film crew, as well as from the few Western tourists who were just beginning to come to Ladakh. The film director observed that, unlike other peoples he had filmed, Ladakhis were unintimidated by the novel equipment or the filming process and projected a sense of equality in relation to their visitors.

When the film crew left, Norberg-Hodge remained with these remarkable people and continued learning their language. Early in her stay, she met a professor from the School of Oriental and African Studies at the University of London, who was impressed by how much Ladakhi she had learned. He encouraged her, and soon she was traveling around the countryside, collecting folk stories and creating a rudimentary Ladakhi-English dictionary as part of a graduate degree program at the University of London.

The trickle of tourists whom Norberg-Hodge had encountered in 1975 was becoming a stream. The Indian government had initiated plans to open up Ladakh, because it was the perfect setting for a tourist industry—an isolated, untainted, traditional culture in starkly beautiful, unspoiled, natural surroundings. Situated in the rain shadow of the Himalayas, Ladakh is a dry land. Talus rubble with scant evidence of life stretches across its valleys and up into the mountains in all directions. Human existence appears tenuous at first in such a barren landscape. Over the past two and a half thousand years, however, the Ladakhis have learned to meticulously divert glacial water at above 11,000 feet through terraced fields of barley and wheat and through garden patches of peas, potatoes, and turnips and at lower elevations into orchards of apricot and walnut trees. Sheep, goats, donkeys, horses, cows, yaks, and *dzo* (a hybrid of a yak and the local cow) provide transport, labor, wool, milk, meat, and dung for fuel. Scattered across the landscape are small villages of several to a hundred houses. Each house is spacious, its ground floor reserved for animals and upper floors used for cooking, entertaining, and sleeping. Walls several feet thick made of mud bricks and stone keep the house livable in the bitterly cold winters. Above the ground floor of some homes, attractive, exterior wooden balconies are attached. Windows and doors have ornately carved lintels. The workmanship and the materials are local, and the design tailored to accommodate Ladakh's environment.

When Norberg-Hodge first walked the paths across Ladakh's 9,000-foot-high valleys and over its 13,000-foot-high passes, she was among the few westerners the villagers had ever seen and the only one who spoke their language. She knew then that she had been blessed with the incredibly rare privilege of experiencing a people whose culture had been virtually untouched by the affairs of the twentieth century, which swirled just beyond the mountaintops that filled their horizon. Norberg-Hodge recalls, "The Ladakhis were free to live according to their own ways and in control of their own lives and their own economy. My experience in Ladakh has really changed my view of the world dramatically—very, very dramatically." Norberg-Hodge had learned that the widespread expressions of human traits such as greed, selfishness, and acquisitiveness in Western culture are not innate but culturally elicited, as are expressions of traits such as cooperation and generosity.

Knowing the Ladakhi language, Norberg-Hodge was able to go beyond superficial observations and experience the intimate character of the Ladakhis and their culture. Prior to coming to Ladakh, she had traveled widely and encountered enough diverse cultures—albeit most of them westernized—to lead her to believe "that human beings were essentially selfish, struggling to compete and survive, and that more cooperative societies were nothing more than utopian dreams." As a product of an essentially industrial society, educated with a specific set of cultural assumptions, she had taken human behaviors in industrial culture as reflections of human nature, the same mistake many of us make because we perceive and assess human behavior, by and large, based on culturally instilled perspectives.

When Norberg-Hodge came to Ladakh, the region was, as it had been for more than 2,000 years, a self-sufficient, loosely connected aggregate of villages spread across 40,000 square miles with a population of a little over 60,000 people. They traded among themselves for the few items that an extended family unit did not provide for itself, and they obtained little from the outside—tea, salt, jewelry, and metals for cooking utensils and simple tools. The cash economy was small, local, and between people who knew each other well. Dishonesty and taking advantage of another person were rare. The household was the hub of economic activity, and women were at its center, thereby giving them considerable status.

Although men and women play different roles in traditional Ladakhi society, these roles complement each other and many tasks are accomplished together. Their language attests to this equality since the same pronoun is used for "he" and "she"—*kho* denotes both. In addition, men and women can have the same name. Cooperation among all members of the community is the standard pattern. At planting and harvest time, neighbors and sometimes the whole community join in—from small children to old people—performing various tasks, often singing and laughing as they work. In the summer when animals are pastured at high elevations, several households will alternate watching over all of the animals. These arrangements that foster cooperation are often formalized in agreed-upon but loosely structured schedules, so changing circumstances, like a sick draft animal or personal difficulties, are accommodated without conflict.

A short growing season of four months, a harsh climate of sun and no rain in the summer, continuous cold often reaching minus 40°F in the winter, and a scarcity of water and arable land—all severely limit Ladakhi exploitation of their land. Many cultural provisions enable traditional Ladakhis to maintain a relatively stable population whose pattern of living respects the fragility of their resources. Each household or extended family has about five acres that are never divided, and this land has to provide for everybody in the family. Polyandry—where a woman has two husbands, often brothers—is common and appears to have played an important role in arresting population growth. Many women do not marry and many of these unmarried women become Buddhist nuns. If a family grows too large for its land to support, usually the youngest son becomes a Buddhist monk. The result of these and other cultural practices is a population whose size is stable and can be supported by the land that it occupies.

Another Ladakhi cultural pattern that comes out of an awareness of finite resources is the efficient use of everything. Essentially every plant, wild or domesticated, is employed for something, and all of a slaughtered animal is utilized. An item is used repeatedly until it can no longer be reused for something else, and even then it is recycled back to earth. Animal and human dung and urine are appropriately processed as fertilizer for the fields. The Ladakhis have no junkyards or waste dumps and minimal pollution. Norberg-Hodge summed it up this way: "Using limited resources in a careful way . . . is frugality in its original meaning of 'fruitfulness': getting more out of little." Ladakhi frugality has enabled them to create an ecologically centered culture that mimics the functioning of long-standing ecosystems—flow energy efficiently, recycle everything, and do not accumulate poisonous waste.

Relations among the people of Ladakh are like those they have with the land. The primary objective is to live well together. Norberg-Hodge found the traditional culture taught people to avoid conflict, and when someone behaved poorly, the offended person usually did nothing. When she asked why, the response was, "What's the point? Anyway, we have to live together." On the rare occasion when an argument or a heated conflict is brewing, the parties talk things over until a solution is found. Customarily a third-party arbiter, often uninvited, comes forward. People seem to want peace over conflict, and they listen to the ad hoc ombudsman, usually accepting the proposed resolution. Even the emotion-laden issue of extramarital sex, although discouraged, is met with the understanding that in human relations these things happen. Losing your temper or getting angry in Ladakh is scorned, as Norberg-Hodge explains. "One of the strongest insults you can hurl at a Ladakhi is *schon chan,* 'one who angers easily.' Angchuk Dawa, a student who has helped me translate folktales, explained that it is bad form for a cuckolded husband to make a scene. 'You see, if he should become enraged and kick up a terrible fuss, it would be his conduct, rather than hers, that would be judged more harshly.'"

Although Ladakhis are deeply religious, their practice of Buddhism is not separate from the rest of their activities. Each house has a chapel, which is the finest room in the home, containing generations of religious artifacts. Prayer flags fly from all rooftops. Every event—from planting to harvest to the full moon to building a house—has religious significance. The people acknowledge the spirits of ancestors and perform rituals to appease them, but Ladakhis seem relaxed, even casual, about their spirituality. Norberg-Hodge's description of the visit of His Holiness the Dalai Lama to Ladakh in 1976 indicates the unity of religion and everyday life: "By the middle of the week-long teaching, the numbers [in Leh, the major city in Ladakh] had swelled to forty thousand. The air was charged with intense devotion, and yet amazingly at the same time there was almost a carnival atmosphere. One minute the man in front of me was lost in reverence, his gaze locked on the Dalai Lama; the next minute he would be laughing at a neighbor's joke; and a while later he seemed to be somewhere else, spinning his prayer wheel almost absentmindedly. During this religious teaching—for many of those present, the most important event of their lifetime—people came and went, laughing and gossiping. There were picnics and everywhere children—playing, running, calling out to each other."

Norberg-Hodge has never experienced a people whose joy of life is so omnipresent in everybody, young and old, almost all of the time. For the past twenty-eight years, she has spent about half of each year among the Ladakhis and concludes, "I have never met people who seem so healthy emotionally, so secure, as the [traditional] Ladakhis. The reasons are, of course, complex and spring from a whole way of life and worldview. But I am sure that the most important factor is the sense that you are a part of something much larger than yourself, that you are inextricably connected to others and to your surroundings."

Their place is intimately known to them, and they resonate in harmony with its seasons, with its constraints. For generations the people have lived in harmony with their place, not in a utopian society but in a respectful arrangement that embraced the ecological limits imposed upon them.

This was the traditional Ladakh that Norberg-Hodge encountered in the mid-1970s, but it was not perfect. Houses were poorly heated with only dried dung for fuel and lacked running water. People died from curable diseases, and infant mortality was higher and life expectancy lower than in many developed countries. Fresh vegetables were not available much of the year, and, even during the growing season, the variety of foods was modest. Electronic communication was nonexistent and contact with the outside world nil. Illiteracy was high because the only formal education was provided to those in monasteries. Modern technologies were just not available. Viewed from the perspective of the developed world, Ladakh was a backward, impoverished place whose people faced daily hardships that no one would bear if he or she could choose otherwise.

Many argue that the developed world has much to offer Ladakhis that would alleviate many of the undesirable aspects of their traditional existence and bring them the benefits provided by modern technologies. At the same time, Ladakh has much to offer the developed world in terms of balanced relations among people and with the environment. Unfortunately, development of places like Ladakh is rarely planned or executed with the well-being of the people or the character of place in mind. Modernization, as currently practiced, is primarily a flow of Western ideas, concepts, and values that erodes the bedrock of an existing culture in fundamental ways. The methods by which change is effected are easily recognized by anyone familiar with developing countries.

In Ladakh, substantive contact with the outside world, which began with the arrival of the tourist industry in the mid-1970s, initiated a cascade of changes that quickly influenced long-established culturally conditioned behaviors. Tight relations in family and community were eroded as demand for new things was created by exposure to tourists, advertisements, and images on film and in print. Young people at first, especially teenage boys, gravitated to Leh in pursuit of what they thought was a better life that they could buy with money. Soon men left their homes to earn the money needed to provide kerosene, wood, or coal for cooking and heating; machines for processing grain; tractors for plowing and cultivation; electricity for lighting and appliances; modern clothes; and innumerable other things that previously had been unneeded or free. Separation between the rich and the poor increased. As dysfunctional families and communities appeared, greed and crime emerged. People lost their self-sufficiency and with it their self-esteem. Women's status began to slip away, because the home-based economy was being replaced by the money economy. Unemployment came into the picture. The population began to grow more rapidly. The culture moved away from the land that had provided Ladakhis with everything to a money economy that depended upon strangers in faraway places.

Norberg-Hodge was horrified as she witnessed this transformation, first in Leh, then in the villages. The Ladakhis had a wholly different perspective, as she relates: "My Ladakhi friends were not thinking of this as destruction. They were curious about things and people from the outside. They were so open, tolerant, unafraid. Their self-confidence made them unable, in the beginning, to realize they would change."

Norberg-Hodge, however, knew all too well from her experience in other places the changes that were brought by industrial culture and its money economy, and she saw the same disruptions emerging in Ladakh. What was beginning to happen in Ladakh, many argued, was the way of progress and unstoppable. Norberg-Hodge, however, did not accept this conventional wisdom. Although she knew of no hard evidence, she persisted in believing that better outcomes were possible if development was accompanied by governmental policies that preserved local character and control, and by a population's understanding of exactly what such development would bring. Then, in 1977, she came across E. F. Schumacher's book *Small Is Beautiful*. In his book, this respected economist says two things: first, the current Western economic system, in the process of delivering immense material wealth to some, ravaged much of what people consider of real value to a community; second, this devastation is not inevitable and can be avoided if an economy is appropriately scaled to a population and adapted to a place. Schumacher's work affirmed for Norberg-Hodge that her contrarian view of development held promise. She firmly resolved that she was "not going to be part of this assumption, which all other Westerners are falling into, that the destruction of Ladakh is inevitable."

Norberg-Hodge is not a Ladakhi, nor will she ever be, but her immersion in their culture provides her with an unusual vantage point from which she can see clearly her own culture. As the Ladakhi people adopted a money-based, market economy with all of its consumer demands, they could no longer meet their own needs and lost the ability to live sustainably on the land. This, Norberg-Hodge saw, mirrored exactly the shortcomings of industrialized cultures everywhere. As early as 1978, she returned to Europe and the United States to bring the message that the current pattern of globalization, with its emphasis on consumption and "free" trade, is a bad idea for most people everywhere, but especially for those in developing countries. She urged others to heed the lessons of Ladakh.

Norberg-Hodge's message is simple: when a people loses access to its traditional resources and no longer practices the ways of life that co-evolved with its landscape over long expanses of time, many desirable aspects of "humanness" that define a healthy community disappear. The example of Ladakh clearly demonstrates that the current pattern of globalization can impoverish people because, by ignoring the unique character and limits of each locality, it erodes environmental stability and produces dysfunctional societies characterized by violence, crime, poverty, substantial disparity between rich and poor, civil unrest, and social insecurity. Norberg-Hodge's response has been to work with Ladakhis, and others, to preserve the best of their traditional culture while embracing those technologies and social arrangements appropriate for their particular environment.

Ladakhis continue to be attracted by the promise of the modern world. They see well-dressed world travelers who spend in an afternoon what a Ladakhi family might spend in a year. These tourists have watches, cameras, radios, computers, cell phones, and fancy travel bags, yet what they do to earn the money to travel and to buy these possessions doesn't sound like work, at least not the physical work that Ladakhis do. Unfortunately, Ladakhis are only seeing the modern world's material successes. The crime, unemployment, drug abuse, inequity, dysfunctional families and communities, physical and mental illnesses, violence, insecurity, stress, and loneliness in industrial countries—all these are invisible to them. Nor do they realize that most people in industrialized societies work more hours than Ladakhis do just to obtain food, shelter, clothes, and other basic needs. Norberg-Hodge gives countless radio interviews in Ladakh and talks constantly with Ladakhis, championing the values of traditional life and presenting a more balanced view of the West. When in Europe and North America, she lectures and conducts seminars on what traditional Ladakhi culture has taught her, highlighting the drawbacks of conventional industrial development and describing an alternative approach.

Knowing that theater was a traditional form of entertainment in Ladakh, in the late 1970s Norberg-Hodge found a way to expand her message. In collaboration with Gyelong Paldan, a Ladakhi who had worked with her on the English-Ladakhi dictionary, she wrote plays; the first, titled *Ladakh, Look before You Leap,* told a more complete story of Western and Ladakhi cultures. Hundreds of people came to see these performed in Leh and other places in Ladakh. Government officials and citizens alike began to talk about the value and character of traditional culture, not only why it should be respected, but also why it should be preserved.

Some of the ways that Norberg-Hodge educates and brings

together the two cultures are more practical. In a land where over 80 percent of the days are sunny, solar energy is the energy source of choice. Norberg-Hodge asked for and was granted government permission to organize a solar technologies project with the primary goal of providing Ladakhis with a better means to heat their homes in the winter. To heat homes they copied a French-designed heat-retaining wall using straw for insulation and local mud bricks painted black to absorb the sun's energy. Villagers also learned how to build solar ovens and greenhouses. In workshops and training sessions, Norberg-Hodge and Ladakhi colleagues brought these innovations to the people of Ladakh.

These projects and activities coalesced to form the international Ladakh Project in 1980. The number of Ladakhis involved mushroomed as sustainable development became an alternative to Western-style globalization that they wanted to explore. Ladakh's leading thinkers were attracted to the work of the Ladakh Project, and they brought their integrity and stature to validate traditional culture. This group was formally recognized in 1983 as the Ladakh Ecological Development Group. In conjunction with the Ladakh Project, this group expanded the venue of appropriate technologies. Now village people cook a variety of foods as well as heat water and their homes with solar technologies; eat greenhouse-grown vegetables in the winter; pump water with gravity-fed, hydraulic pumps; and provide mechanical power with water turbines that improve upon traditional water mills. Most important, the people who use and benefit from the new technologies have been intimately involved with their utilization from the beginning and are responsible for operating and maintaining these new technologies.

The Ladakh Ecological Development Group's headquarters, the Center for Ecological Development, was built in Leh, not only to provide a central location, but also as a "presence" available to Ladakhis, government officials, and visitors. The center's architecture follows traditional Ladakhi design, with a restaurant that serves local food cooked in solar ovens. The center's library highlights ecological issues and sustainable development. A handicrafts area features the work of local weavers, silversmiths, woodcarvers, tailors, embroiderers, and painters. A variety of workshops, conferences, and educational programs are held at the center. This is a place where visitors and Ladakhis come together to learn from each other, and two similar centers have been built outside of Leh.

In 1991 the Ladakh Project became the International Society for Ecology and Culture (ISEC) with offices in Ladakh, Britain, and the United States. In that same year Sierra Club Books published *Ancient Futures: Learning from Ladakh,* in which Norberg-Hodge tells her story of Ladakh. Since then, this book has been translated into more than thirty languages, and the universal response from people in developing countries everywhere is: "This is our story, too." For over twenty years, ISEC has sponsored "Reality Tours" that bring influential people from Ladakh to Europe and North America, and bring westerners to Ladakh so members of both groups can observe firsthand each other's cultures. Each year over 3,000 tourists who come to Ladakh participate in ISEC's tourist education program, which seeks to instill respect for Ladakh's culture and to stimulate tourists to work for change in their home communities.

In 1995, ISEC organized the Women's Alliance to provide Ladakhi women with a community-based organization whose explicit agenda is to preserve and celebrate traditional culture and agriculture. Within five years the alliance had over 4,000 members and a presence in almost a hundred villages. The group has held workshops and festivals celebrating traditional food and farming, sponsored "no TV" weeks and bans on plastic packaging, and initiated a program to protect cultivated indigenous plants. Its success is reflected in the words of Dukchen Rinpoche, the head of Hemis Monastery in Ladakh, who told the Women's Alliance, "Your message about protecting Ladakh's culture and spiritual foundations is one that I've tried to get out many times, but you are the ones who have actually enabled it to take root."

ISEC and all of its programs are efforts to address a crisis of perception. Norberg-Hodge puts it this way: "What ultimately feeds us and makes the world go around is the living world itself. When we discard that understanding and act as though it is of no importance and follow the laws of the human-made system, we rapidly strangle all life and in the process make ourselves much less happy."

Many data and analyses authenticate Norberg-Hodge's assessment that humanity faces a crisis of perception and her concern over our misplaced faith in the current global economic system, which fails to accept ecological limitations. Perhaps the most sobering picture is provided by ecological footprinting, a technique for assessing the impact of a particular lifestyle on the earth's resources. Although political and corporate leaders, and many others, claim that the globalization of trade will alleviate poverty and enable everybody to achieve the lifestyle of those in the Western industrialized world, footprinting tells another story: if we want to achieve such standards of living for all, we have no choice but to substantially increase the ongoing consumption and pollution of Earth's ecological resources— forests, prairies, wetlands, soil fertility, wilderness, fish stocks, wildlife, oceans, atmosphere, aquifers, among others. The resulting magnitude of loss of ecological resources would mean a severely accelerated reduction of Earth's life-support capacity, which in turn would precipitate a mass die-off of life, including humans.

We don't have to wait, however, for more globalization of the type currently advocated by politicians and corporate leaders for Earth's life-support capacity to be eroded. Ecological footprinting tells us that we are currently using about 30 percent more ecological productivity and assimilative capacity than the earth can provide on an ongoing basis. That is, we are already consuming natural resources faster than they can be replaced and overwhelming the assimilative capacity of nature. Increasing consumption around the world will only exacerbate an already dangerous trend. More fundamental, we must question the desirability, let alone the possibility, of the objective of raising everyone's level of consumption to that of industrialized societies. There is little evidence that bringing such a materialistic lifestyle to people of traditional cultures has positive results for anyone but the owners of human-created capital and natural resources who champion the current pattern of globalization.

With the modernization of Ladakh have come the physical illnesses associated with industrialized Western civilization—diabetes, heart disease, cancer, and stroke. Since many of these are occuring more frequently than before or are new diseases for which there are no effective traditional remedies, there is an increased need for Western medicine. Some drugs, such as antibiotics and other scientifically based treatments of common ailments, have been beneficial. And procedures for dealing with the new illnesses associated with modern civilization have helped many Ladakhis. However, when Western medicine is introduced to treat new diseases, it is also used to heal traditional illnesses and thereby displaces traditional medicines. Wholesale replacement of traditional remedies by Western medicine is not necessarily beneficial since not all effective traditional treatments are replaced by modern ones, nor is access to Western medicine universal (in contrast to traditional treatments that were available to the entire community). As the demand for traditional remedies diminishes, fewer traditional healers are trained to find and prepare medicinal herbs and to treat patients. This indigenous, ancient knowledge is then lost, and what was once free and provided locally by nature must be purchased from distant sources.

The traditional Ladakh house was built on family land with local labor and from local resources: stones, mud, and wood. People worked together and little if any money was exchanged. As people began to crowd into Leh, seeking money and a better life, land became a commodity, stones and mud were not available for the taking, and labor was no longer free. Since making mud bricks is a slow process and the bricks required transport into Leh, subsidized cement and steel imported over the Himalayan mountains were less expensive. In addition, since working with steel and cement took special training, Ladakhis could no longer build each other's houses. They had lost house-building skills and the cooperative attitude of their traditional culture. Skills and technologies that were once local, cost nothing, and nurtured the community had slipped away.

Modernization associated with migration to Leh and industrial agriculture has also undermined local agriculture. Wheat flour imported into Leh from other parts of India is less expensive than flour from local villages. "As a result," Norberg-Hodge notes, "it becomes 'uneconomical' to grow your own food—an unimaginable concept within the traditional economy." It seems contradictory that food traditionally grown locally costs more than the same crops grown, harvested, packaged, and transported from hundreds of miles away. This situation exists, on an even larger scale, in other parts of the world.

Tomatoes from Israel are sold in New York City in August when local farms are producing delicious tomatoes. In 1996 Britain exported 103 million pounds of butter while importing 108 million pounds of butter. The same holds true for milk: 117 million quarts exported from Britain and 183 million quarts imported. The economist Herman Daly observed, "Americans import Danish sugar cookies and Danes import American sugar cookies. Exchanging recipes would surely be more efficient." So why don't we just swap recipes and feed ourselves locally, except for some special items that cannot be grown close to where we live? The situation is complex; nevertheless, some light can be shed by considering the origins of the globalized economy and the degree to which it is subsidized.

At the 1944 Bretton Woods conference in New Hampshire, two international organizations were established: the International Monetary Fund (IMF) and the World Bank. In 1948 the United Nations created an important international system, the General Agreement on Tariffs and Trade (GATT), which would facilitate the functions of the two other organizations. The World Bank funds giant infrastructure projects, such as centralized energy facilities, long-distance transport networks, and communications systems. The IMF implements and oversees a growth-based, economic system for all national economies, while GATT facilitates international trade by removing tariffs and other barriers to "free" trade, which may be defined as the movement of goods across national boundaries without restriction or tariffs. In 1994 the GATT nations formed the World Trade Organization (WTO) to set trade rules and arbitrate disputes, with the objective of making national economies better suited for foreign trade and investment.

These international organizations have nurtured a large-scale industrial agriculture that has led to a substantial reduction in local food production. Since the 1950s the number of farmers in the United States and Europe has decreased by over 60 percent. In developing countries like China, people living off the land decreased from 92 percent to 40 percent since 1979. Each year in the United States the distance that the average food item travels before it arrives on the grocery shelf increases—over 1,000 miles in 2000. In the United Kingdom the distance traveled increased by 50 percent in the last twenty years. World trade in food increased from 200 million metric tons to over 600 million metric tons from 1965 to 1998. In industrialized nations, food is primarily controlled by a few transnational agribusinesses that control production from "seedling to supermarket." For example, ConAgra, Inc., accounts for six cents of every food dollar spent in the United States; Philip Morris is the leader at ten cents on the dollar.

On one hand, globalization of industrial agriculture, in tandem with government subsidies, has succeeded in keeping the direct cost of food to many consumers low and increasing the number of people that each farmer can feed. In the United States the price of a typical grocery cart of food has increased only 3 percent in about twenty years. On the other hand, the domestic farmer's return, about 30 cents on the dollar from 1970 to 1990, decreased to about 20 cents in 2000, or by 33 percent. In addition, industrial agriculture has been disastrous for the environment and led to the impoverishment of rural communities. And this agriculture is phenomenally energy inefficient. It takes about 3 calories of fossil fuel energy to produce 1 calorie of food on the farm and another 10 or so calories to get this same calorie of food to the store—like spending $13 to earn $1. And large-scale agriculture and the low prices it delivers to the consumer have not managed to end worldwide hunger: 800 million people were chronically malnourished while 2 billion were hungry some of the time in 1999.

The standard perception is that globalization is driven by the "free" market, a hypothetical construct of economists whereby goods and services are exchanged in the most efficient way to ensure that resources are correctly priced. In theory one might be able to create a market system in which the full cost of the food exchanged is reflected in the price. In practice, however, this is not possible, except perhaps locally, because only local food economies can do without all the physical and administrative infrastructure provided by the IMF, GATT, the WTO, and the World Bank. When this infrastructure is supplied essentially free to large, industrialized farms and corporations, it puts small local farms at a financial disadvantage. In addition, pervasive subsidies exist because each nation seeks advantage for its products internationally as does each company within a nation. Conservative calculations show that the world is spending annually at least $700 billion to subsidize agriculture, energy, road transportation, and water. These types of subsidies in the United States and European countries enable them to export and sell crops 20 to 50 percent below production cost, thereby pushing global prices down and putting local farmers in other countries at a disadvantage.

Subsidies keep the price of a commodity below what it would have been without the subsidy and can, within the context of market transactions, appear to benefit both consumer and producer. In a wider context, however, as the story of Ladakh shows, international trade, and the subsidies it elicits, may have substantial negative consequences, not only for the environment, but also for many people. The negative consequences are not usually immediate, nor are they readily linked to subsidies. In addition, globalization handsomely rewards people holding positions of power and influence in corporations and government, who out of self-interest perpetuate the process rather than act on the basis of objective evaluations. And, equally important, the general public has been culturally conditioned to accept globalization, not only as a result of natural processes and therefore inevitable, but also as basically good.

Norberg-Hodge minces no words in challenging the assertion that the present pattern of globalization feeds the poor and provides them with a decent lifestyle: "It's a lie." She has seen that it is not the answer, but her question remains: Can the industrial, consumption-based, political economy engulfing the planet be transformed into a more benign process? After almost twenty-five years of questioning the present system, Norberg-Hodge sees a growing awareness among people everywhere that the current economic system will not benefit the bulk of humanity or the environment. She considers the upwelling of protest against the WTO in Seattle, Washington, in 1999 as the turning point. In those protests, organizations from all over the world— including Movement for the Survival of the Ogoni People, Indigenous Peoples Coalition against Biopiracy, Rainforest Action Network, Third World Network, International Forum on Globalization, Transnational Institute, and hundreds more—rallied to shut down the WTO meeting. Their names are unrecognized by most of us, but these groups represented steelworkers; farmers; teachers; scientists; students; indigenous people; activists for human rights, labor, and social justice; and environmentalists of all stripes who raised their voices against the corporate agenda of "progress" imposed on them. When Norberg-Hodge, a founder of the International Forum on Globalization, spoke at a concurrent counterconference to the WTO meeting in Seattle's symphony hall, every one of the hall's 2,500 seats was occupied and about 2,000 people had been turned away—interest in challenging the current pattern of globalization was so strong that $10 seats were selling for $40 on the street.

The thousands of protesters who shut down the 1999 WTO meeting in Seattle represent but the tip of the iceberg. A growing number of the world's people are unwilling to quietly accept the private decisions of government-sanctioned corporations and organizations like Monsanto, General Electric, the World Bank, and the WTO, because these actions determine the character of their lives and the fate of their local environments. Ladakhi citizens did not participate in the decree that brought tourism and industrial culture to their villages and homes. However, they and myriad others were in Seattle to ask for their place at the decision-making table. The U.S. and other elected governments champion the ideas of freedom and democracy while wholeheartedly supporting corporations and organizations that operate in private, without public input, scrutiny, or accountability to determine the fate of all humanity.

Since Seattle the meetings of every international trade and financial organization have been vigorously protested. The schism between those that have power and those who seek to have their voices heard by those in power has led to polemics and belittling accusations. Norberg-Hodge cautions: "We need to move away from blaming and towards understanding. We are going to get out of this mess much better if we move beyond demonizing anybody. That includes heads of corporations, politicians, and a whole range of people who have played quite an active role in promoting the continuation of this system. From my point of view these people have been indoctrinated into a belief about how things work. Their belief is that, above all, the economy makes the world go around. It has to be kept robustly alive, because if the economy suffers, then people will suffer and so will nature. As a consequence we are caught in this situation where our leaders have pushed the economy at the expense of everything else. They see greenies as very well intentioned and understandably caring but also as naive and limited, without real understanding. So this allows a lot of our government and business leaders to quite ruthlessly undermine the green movement, because they believe they understand what is in the best interest of the planet."

This deep belief and faith in the growth economy is publicly displayed every day. In the United States, almost immediately after the terrorist acts of September 11, 2001, George W. Bush urged people to get back on airplanes and spend to support the economy and our way of life. A group of people flew from the West Coast to New York City to spend, as a group, a million dollars to support the economy. A woman in her 2001 Christmas letter to my wife and me proudly announced that to support the economy she had cleaned out her closets and then gone on a spending spree to replace what she had tossed. In an even more egregious example of giving priority to economic growth over environmental health, the Bush administration in early June 2002 in a report to the United Nations concurred with the United States National Academy of Sciences and the United Nations Intergovernmental Panel on Climate Change that human activities, namely burning fossil fuels and cutting down forests, are causing climate change, which is likely to have disastrous consequences for the United States and the world. However, the administration has refused to take direct action to curtail carbon dioxide emissions, insisting that such actions would disrupt economic growth.

Growth and consumption are considered by governments and corporations as the panacea for social and environmental ills alike. As Norberg-Hodge notes, "Our governments are wedded to a notion of growth that means they use our taxes to promote consumerism. So even what is happening in the universities, on television, and in our schools is being transformed and shaped to encourage overconsumption. Overconsumption is fairly easy to define when the average person in the United States uses many times his fair share of resources. Just eating a cup of yogurt provided by the global commercial system results in consuming substantial amounts of petroleum in the form of plastic, transport, refrigeration, and all sorts of things that are unnecessary. It would not be difficult for us to get out of this situation if we understood the systemic linkages that led to overconsumption. Our Roots of Change program works with small community groups that sit down and talk about these issues to effect change."

The core of ISEC's Roots of Change educational program is small, self-run groups that use ISEC materials to address the problems associated with globalization by establishing the value of local economies to the health of the community and the importance of not joining the consumer culture. Beginning in 1991, dozens of groups in the United States, the United Kingdom, and elsewhere have completed the program and gone on to start farmers' markets, community newsletters, and other local community endeavors to counter globalization and consumerism. In contrast, most education is geared to preparing students to be consumers in a growth economy rather than teaching them how to reduce consumption. Globally some $450 billion, or $70 per person, were spent on advertising in 2002 to convince people to consume.

Although the tidal wave of a consumption-based globalization is not abating, Norberg-Hodge is more optimistic about its changing than she has been in almost three decades. The protests in Seattle against the WTO and other similar protests in Washington, D.C., Genoa, Ottawa, Cancún, and most recently in Miami against the proposed Free Trade Area of the Americas showed her not only that worldwide people desire to participate in the decisions their governments make, but also that people in urban areas throughout the world are awakening to the value of community and nature. She observes that "in the heart of Western urban culture, where people have lost community, family, and contact with nature, some individuals have nevertheless discovered the joy of connecting to the natural world, caring for animals, and nurturing other people. I see this culture that separates us from the living world and from community as alienating. But, very inspirationally, when people have been deprived of and separated from nature, they often find their way back. Where do we see the strongest environmentalism? It is in the big cities of the world: Barcelona and Madrid in Spain, New Dehli and Bombay in India, New York City and San Francisco in the United States. This conscious appreciation of the living world as opposed to the static human-made world is a very important thing. Universally I believe deprivation engenders a desire to get back to the living world."

Interestingly, Norberg-Hodge finds a split between the environmentalism among urbanized populations just described and the desire for development among rural-based populations of agriculturists and nomads who have never left the land. The rural populations do not often practice a conscious environmentalism, nor are they aware of the need to accept slowly and selectively industrial development, which is often perceived as unqualifiedly beneficial. She sees this split wherever she travels; it transcends geographical and national boundaries. The distinction is seen between the people of Leh and those in Ladakh's villages. Norberg-Hodge's programs in Ladakh and elsewhere attempt to facilitate a closure of this gap between rural and urban people by bringing them together. Urban dwellers require the knowledge of rural people to reconnect to the land effectively and thereby create more locally based lives, whereas country dwellers can use the experiences of urban people to resist wholesale industrialization by selecting those technologies appropriate for their particular places.

Rural people who have never left the land usually produce much of their own food, whereas most urbanites, particularly in industrialized societies, buy their food in a store, much of which has been produced by industrial farms in distant locations. For city dwellers, growing food in a garden or buying food grown by small local producers holds great promise for decreasing the scale of globalized trade of food and for reconnecting people in urban areas everywhere back to the earth.

I grew up in a small town in northern New Jersey and have always lived in an urban setting. My dad had a victory garden during World War II, and he continued to grow some of our food until we moved in the mid 1950s to a house just outside of Milwaukee, Wisconsin, that had a postage stamp–sized yard. As a youngster in New Jersey, I had a little garden of my own, and as an adult I have followed my father's tradition. We have had a vegetable garden ever since I broke sod behind the garage of our Naval Station housing when, in 1969, I taught at the Naval Academy in Annapolis, Maryland.

In the beginning our family garden, like the early gardens at The Land Institute, had an economic basis. Now, the garden is a gesture toward a more ecologically connected and healthy way of living. At the local supermarket the carrots, broccoli, squash, tomatoes, potatoes, corn, and lettuce travel on average about a thousand miles before we pick them off the shelf, and they consume in production and transportation about thirteen calories of energy for every calorie of energy they deliver to us at our dinner table. Most have been sprayed with toxins, and the soil in which they have grown has eroded. In addition, the people who handled them, as well as the farmer who grew them, were often paid little more than the minimum wage.

Having a garden is a step toward a more ecologically durable and socially just society with tangible personal rewards. Garden-fresh produce is more nutritious, less toxic, and tastes better than store-bought food. I enjoy a garden-fresh tomato but have little interest in eating one that tastes like cardboard because it has been bred for long-distance travel and to be harvested weeks before it is eaten. Being outside working the soil, planting the seeds or seedlings, watching the plants grow, harvesting the vegetables just before eating them, and even weeding—all connect me to the biological world of which we humans are a part. Too often we forget that our deep evolutionary roots are not in shopping malls, cars, airplanes, houses, or other human environments devoid of Earth's organic diversity.

As Edward O. Wilson contends in his book *Biophilia,* humans have an innate tendency toward what he calls "biophilia," our fascination with and love of other living organisms. We go to great expense and effort to have and keep pets. We bring plants into our homes and landscape our yards, and many of us like to visit zoos or botanical gardens, or go hiking and camping. The biologically rich settings and vistas we tend to seek out relate to our evolutionary past. We gravitate to open spaces with a few trees and a body of water, areas similar to the African savanna. It also appears true that, like other latent capacities, biophilia is enhanced and nurtured when one is exposed to nature early in one's life. Predictably perhaps, there seems to be a strong correlation between early, in-depth exposure to nature and other living organisms and strong avocational and vocational commitments to studying the life sciences as well as advocating for the environment. The reverse also appears to be true: the capacity for biophilia can wither. With people in general and children in particular, the cocoon of the human-mediated world, which isolates us from our evolutionary habitat of intimate contact with biological diversity, may diminish biophilia in modern cultures.

Beyond the personal connections and satisfactions, on a small scale gardening addresses other problems. Composting leaves and other yard material, along with kitchen plant waste, builds soil and provides fertilizer. Mulching plants with grass cuttings eliminates most weeds and retains soil moisture while they decay to enrich and create soil. In a small garden, manual removal of insect pests like Colorado potato beetles and tomato horn worms is often sufficient to keep pests at bay, thereby averting the use of toxic chemicals. Organic gardeners also have ecologically based ways to mimic complex ecosystem processes to the gardener's advantage.

Troy, New York, like some other American cities, has had for almost three decades a community gardens program so that its citizens in apartments or homes without appropriate space can raise some of their own food. Abandoned and undeveloped lots are rented or purchased and then converted into gardens. For a small fee a person can have his or her own growing space. It may not seem like much, but if everybody had a garden, the percent of locally grown food would sky-rocket. In our garden, vegetables grown in a modest 10-by-70 foot space routinely provide most of our vegetables in the growing season, and we have tomatoes, broccoli, and Swiss chard into December or January. When we have an especially good harvest, our carrots, beets, and squash last into the next spring. As a young couple, my wife and I spent time canning and freezing vegetables, resulting in our having food from the garden into the next growing season.

ISEC champions local agriculture as essential in the effort to connect people with the land and redirect globalized trade. Two ISEC publications present the case against globalized trade in food and the case for local food economies—*From the Ground Up: Rethinking Industrial Agriculture* and *Bringing the Food Economy Home: The Social, Ecological and Economic Benefits of Local Food*. ISEC implements change by working with numerous groups in the United States and Europe to promote the eating of locally grown food. England had no farmers' markets in the mid 1990s; however, over 270 were established by 2000. In England today, over 40,000 consumers are connected to local farmers through such markets. Farmers' markets in the United States totaled over 3,000 in 2002, an increase of 75 percent since 1994. Farmers are being linked to consumers in a mutually beneficial producer-consumer unit called "community-supported agriculture," generically referred to by the acronym CSA. In such arrangements consumers are provided with a pre-established fraction of the farm's yield in return for an up-front payment to the farmer and in some cases a lower cost in exchange for helping on the farm or assisting in the delivery of the produce to other consumers in the CSA group. The practice began in Switzerland some thirty years ago and is now global. CSAs can be found in most areas in the United States; in fact, we have several in the Capital District area of New York where I live.

These local food initiatives are heartening, but are they capable of replacing large-scale industrial agriculture? Is it possible to provide most of the food for the majority of the world's 6.4 billion people, and the several billion more on the way, with an agriculture of family gardens, farmers' markets, CSAs, and store-sold locally grown food—all produced by a Jackson-type natural systems agriculture? Along with many others, Lester Brown, founder of Worldwatch Institute and president of the Earth Policy Institute, is skeptical that the current global food system can be replaced, by and large, by local agriculture. Cuba, however, has been running an experiment that encourages me and others.

With the breakup of the Soviet Union, the coalition of European communist countries fell apart in 1989, leaving Cuba in trouble. At that time Cuba imported most of her fuel, two-thirds of her food, and four-fifths of her machinery. In fact, 85 percent of her trade was with European communist countries. Immediately, with communism's collapse Cuba's purchasing capacity dropped by 60 percent. The long-standing, economic blockade by the United States hindered trade with other parts of the world. A desperate situation was further exacerbated because food aid and other assistance was not forthcoming from the international community, which is unusual in such a situation. The United States tightened the blockade with the Torricelli bill—the Cuban Democracy Act—in 1992, which banned shipment of food and medical supplies to Cuba, and the Helms-Burton Act of 1996, which further restricted investment in that country. As a result, Cuba was isolated and had to solve her food crisis internally. Since massive food aid historically has not helped afflicted people to provide for themselves in the future and often increases their dependence on food imports, in the long run these shortages turned out to be a blessing in disguise because they forced Cubans to produce their own food.

At first, yields dropped as imported pesticides, fertilizer, and fuel disappeared. The prevailing tenets of industrial agriculture initially led Cubans to launch a major effort to replace these yield-limiting inputs of industrial agriculture with biologically produced pesticides and fertilizers. The intensity of the crisis and Cuba's agricultural history, however, helped Cubans to understand that their problems were systemic and could not be solved simply by substituting local inputs for imported inputs. They realized that large-scale, industrial monoculture—even with biological rather than chemically based fertilizers, biocides, and energy sources—was still unsustainable because it was not solidly grounded in the principles of ecology and evolution.

Cubans embraced highly diversified, small-scale farms. They broke many of the large state farms, which had used over half of the island's cultivated land, into smaller farms. The transition from industrial agriculture to complex, ecologically based food production was facilitated by widely implementing traditional farming practices that still persisted in rural areas and by using alternative approaches to fossil fuel–based agriculture, which had been developed following the oil shortages in the 1970s. With subsidized machines and chemicals gone, small farms were found to be more efficient in terms of overall yield than the large state farms they had replaced.

The removal of subsidies and the acute food shortages of the early 1990s had caused prices to increase. Urban gardens appeared everywhere, and they were nurtured with technical support from the government. Backyards and vacant lots quickly became sources of eggs, pork, and all kinds of vegetables that undersold similar items from large farms while still providing a profit. With three-quarters of Cubans living in urban areas, small urban farms and home gardens became an important component of food production. Within half a dozen years urban agriculture had created 160,000 jobs associated with growing and marketing produce, and in 1999 almost a million tons of food were grown in urban settings.

Cuba is no longer in a food crisis. In 1998 the production of some items like vegetables and beans exceeded that of 1989, whereas that of others like milk, eggs, and beef were about half pre-crisis levels. Cubans did not voluntarily abandon industrial agriculture—they had no choice. Since they no longer face major food shortages and their economy has recovered, it remains to be seen whether they will persist to complete the transition to a Wes Jackson–like natural systems agriculture. But we now know that a society with remnants of traditional agricultural knowledge and a high level of technological, scientific competence can abandon industrial agriculture and change quickly to a local, ecologically based agriculture.

Urban gardens, farmers' markets, and community-supported agriculture combined with ecologically based, regional farms provide a model of agriculture that can adequately feed people while supporting local communities. Will these admirable examples, in conjunction with the awakening Helena Norberg-Hodge sees in people around the world, enable us to adopt the approaches to local living that have emerged from the margins? Can we structure our social institutions and relations with the environment in ways that are compatible with our species' evolutionary roots as a small-group, place-based social animal while continuing to establish dynamic, creative, and peaceful global associations with each other? It is vitally important that we begin to do so. We now exist in a globally connected world, but the health of the planet and our communities depends on our ability to live on a human scale, within local economies that are sensitive to the place-based habitats of which we are a part.

SIX

Be Fruitful and Few

Werner Fornos and Population

> Slowing down population growth in culturally sensitive and reli-
> giously respectful ways has become a priority that all of us have to
> address because there are no acceptable humane alternatives.
>
> WERNER FORNOS

The population numbers tell the story of the human species' spectacular biological success. A few hundred thousand years ago, when our
species was young, we were thousands strong. With the advent of agriculture in the Middle East some 10,000 years ago, our distribution had
become global, and our numbers had grown to several million. It took
almost all of the time from our origins as a species until the beginning
of the nineteenth century for the human population to go higher than
1 billion. Then in a mere 200 years we added 5 billion more—the last
billion in just 13 years! In 2004 some 6.4 billion of us inhabit Earth,
with several billion more predicted before the middle of the century.

Evolution works as it does because all life is endowed with this
biotic potential—the capacity for exponential population growth. Yet
the dynamic tensions among organisms within their ecological niches
constrain extravagant success before it happens or crush an organism's
population when its size grows out of control. Evolution has provided
humans with the ability to assess the present and see a future of no
population growth or of negative population growth as a result of biological and physical constraints. Aware that Earth's resources cannot
continue to support exponential human population growth, we ask:
first, how many people can Earth's resources support and for how
long? second, what will happen when our numbers exceed this limit?

Many have tried to calculate the maximum number of people
Earth can support at one time, leaving aside any consideration of

duration. At the current growth rate, the human population of about
6 billion is some 50 years away from doubling, and 12 billion people is
the estimated mean number of people that Earth can support, a number arrived at by those who base their calculations on realistic biological and physical parameters. Twelve billion people, however, is most
assuredly more than Earth's carrying capacity for humans—the number it can support on a continuing basis.

The question "what will happen when the carrying capacity is
exceeded?" has become critically important for those concerned with
formulating global policies to avoid catastrophes caused by overpopulation. If we go beyond this limit—intentionally or because we have
overestimated it—it is highly likely that major elements of life-support
systems on the planet will be impoverished, causing a collapse of the
population. Scientists have warned of this possible crash for several
hundred years with many scenarios of how it will be manifested. All
project that nature's correction will be catastrophic. Thus, many have
advocated that humanity take things into its own hands before this crisis happens.

Paul Ehrlich's *The Population Bomb*, published in 1968, focused
attention on the dire consequences of continued population growth—
disease, starvation, resource shortages, infanticide, civil strife, war—as
Thomas Malthus's "Essay on the Principle of Population" had done
almost two centuries earlier. In the 1960s many were concerned that
the postwar baby boom in the United States, if persistent, would push
the country's population over 400 million in the second decade of the
twenty-first century, thereby causing resource shortages and environmental degradation. President Richard Nixon in 1969 said population
growth is "one of the most serious challenges to human destiny in the
last third of this century." Consistent with this assessment, he supported legislation that created the Commission on Population and the
American Future, chaired by John D. Rockefeller III.

In 1972, after two years of study, the Rockefeller Commission
arrived at the consensus that population growth is not essential to the
U.S. economy: "the health of our country does not depend on popu-

lation growth, nor does the vitality of business, nor the welfare of the average person." To arrest population growth before natural resources were overexploited, it proposed to decriminalize abortion, remove any remaining legal barriers to making contraceptives freely available, provide sex education for everybody in schools and other community institutions, eliminate all sex-based discrimination, restrict annual immigration to 400,000, and block illegal immigration. The commission proclaimed that, with the knowledge and means available, couples will freely choose a family size appropriate for them and the country. Political pressures mounted against the commission's proposals, especially from large religious groups and conservative organizations, while at the same time fertility rates were beginning to decline in the United States. Nixon lost his fervor for overtly addressing population growth and rejected the commission's report. However, many in government and throughout society believed the commission was right and persisted in implementing its recommendations.

One elected official who held course was Werner Fornos, a German immigrant. In 1966, at thirty-three, he had been elected to Maryland's state legislature. In committee hearings Fornos listened to witness after witness claim that women, especially poverty-stricken women, were having children because they wanted bigger welfare checks. Many of his fellow legislators agreed, but where were the facts to support such claims? Had these claims been substantiated? Since Fornos could not find definitive research on the subject, he set out to prove or disprove the merits of the argument. His investigation showed that the claims were unsubstantiated: many pregnancies were unintended and were in fact the result of women's inability to regulate their fertility. Women got pregnant for myriad reasons, but not to receive government handouts. Fornos thought that to perpetuate this fallacy and not provide people with the knowledge and means to practice family planning was an egregious assault on women in particular and, more broadly, on our deeply avowed principles of justice and equality for all. Fornos had found his cause.

～

Werner Fornos was born Werner Horst Fahrenhold in Leipzig, Germany, on November 5, 1933. On December 3, 1943, his life suddenly changed when the apartment building where he lived with his mother and five siblings was destroyed by Allied bombs. Buried in the rubble he tapped on water pipes for three days until a search party found him, but he was rescued only to find his family gone. He would not learn until the mid 1950s that his mother and siblings had survived; his father, a soldier in the army, was killed in 1944. By the time he found his family, his life and theirs were worlds apart. On the night of the attack Werner and several friends had been caught on the roof of a local church, setting off firecrackers. They were charged by a local Nazi interrogator with treason for helping the Allies to find the city. For this reason, not long after Werner was rescued, he and the other boys were sent to a reform school where punishments were severe. As a penalty for reading a magazine without permission, for example, Werner was whipped and given solitary confinement in an unlit cell with only bread and water for sustenance.

In July 1944 Werner managed to hide in the brake box of a troop train and escape to Normandy, France, only to find himself in the midst of a battle. By chance, members of the Twenty-ninth Infantry Division of the United States Army found the ten-year-old boy after he was dazed by an artillery explosion. Werner emotionally recalls what happened: "The most important contributor to who I am is really a larger family. Not having had any real contact or access to my natural family until I was over twenty years old, I found that larger family was the GIs who adopted me in World War Two. They taught me how to read and write English. Taught me values. I was treated as an equal even though I was a young boy who needed their help more than anything. If it wasn't for them, I wouldn't be here today. So the real underpinnings in my life come from many of the soldiers of the Twenty-ninth Infantry Division and subsequent military units who took me in and cared for me." But the help was not one sided: he assisted the U.S. soldiers in finding food and would cross front lines to spot enemy positions.

When the war ended, Werner knew he wanted to leave Germany and go to the United States. His army family hid him on the ship that was to bring them home; however, he was discovered and placed in the hands of the German welfare system. Almost immediately, he left his German caretakers and stowed away on another troop ship, only to be returned to Germany, this time from Brooklyn, New York. His next attempt at running away brought him back to Brooklyn, but again he was sent home. Three was almost a "charm:" he got as far as the address in Ohio given him by an American corporal who had said to look him up if Werner ever got to the United States, but the corporal had moved. Werner recalls, "Out of money and hungry, I entered a short story contest held by the local ladies auxiliary. I won fifty dollars, but the telling of my tale led to my being turned over to immigration officials."

Once again in Germany, Werner was sent to an area under Soviet control. As he spoke English well and was carrying a U.S. Army handbook, he was held and questioned for five days by Soviet administrators. The verdict: the boy was a spy. His life sentence started at the uranium mines near Annaberg, Germany, working in the mine bakery. Within three weeks, he escaped with the help of coworkers by hiding in one of the large breadbaskets used to deliver bread to guard posts surrounding the camp. Traveling on foot by night, he eventually made it to the British zone in Germany. He got a job as an interpreter for the British High Command and there learned that the Soviets were looking for him. On the move again, he entered the zone controlled by the United States. Washing cars for his keep, he made his way to the Rhein-Main air base. Working three jobs—as an interpreter in the day, a pinboy at the base bowling alley in the evening, and when needed a worker loading planes for the Berlin airlift—he planned his fourth trip to the United States. On January 4, 1950, he stowed away on a C-54 transport plane being flown to the United States for overhaul. Dressed in jeans and a jacket ordered from Sears, Roebuck, Werner pretended he was the son of the pilot when the plane arrived at Westover Air Force Base in Massachusetts. However, an officer from

Rhein-Main air base recognized Werner and turned him over to the authorities.

He was sent to East Boston Immigration Detention Station for deportation. Not long after Werner arrived, members of the local Newton, Massachusetts, Congregational Church came to visit those awaiting deportation. After hearing his story, Lilly Fornos, a member of the group and an immigrant herself, obtained permission to have him stay with her family until deportation was arranged. Immediately, the Fornoses orchestrated an all-out effort to have him made a legal immigrant. The efforts were successful, and in due time Werner took Fornos as his last name.

Ever since he had gained legal residency in the United States, Fornos had wanted to devote himself to public service. He had a gift for working with people to get things done, and he engaged in constant activity. Fornos describes his schedule as a teenager: "I went to work as a dishwasher while in summer school. I went on to become a copyboy at the *Boston Post,* working after school from four in the afternoon until one in the morning. Once the paper was put to bed I was able to do my homework with not as frequent interruptions." Work and other life experiences, not school, were Fornos's classroom. With only a summer school session in math, the sixteen-year-old Fornos was tested and placed in tenth grade in Newton, Massachusetts, just one year behind for his age after having not been in a classroom since he was ten. On completing high school, he entered Boston University, but after one semester he was drafted into the U.S. Army. Upon leaving the army he held a number of government positions, including ones in the Kennedy and Johnson administrations. All the while he attended night school at several universities, eventually graduating from the University of Maryland in 1965 with a bachelor of arts degree in political science.

After four years in the Maryland legislature from 1966 to 1970, he became assistant secretary of human resources as well as manpower administrator for the state of Maryland and made his first bid, unsuccessfully, for a seat in the United States House of Representatives. He left the Maryland state positions after two years to become executive director of Planned Parenthood for metropolitan Washington, D.C. Four years later he became director of the Population Information Program at George Washington University in Washington, D.C., and made his second run for the House of Representatives. Defeated again, he accepted that public office was not how he would serve. All the while, population concerns dominated Fornos's life—in addition to his full-time job, he was an independent management consultant and evaluated family planning programs in Bangladesh, China, Indonesia, Mexico, and other countries for clients such as the United States Agency for International Development (USAID).

During this time Fornos noted the many emerging programs that focused on domestic population issues, including the Pregnancy Advisory Service, later to become the Population Institute, founded by Rodney Shaw, a Methodist minister. Fornos was also keenly aware that the population in developing countries was exploding just as predicted by Paul Ehrlich—the world population was growing by over 70 million each year, with more than 90 percent of that growth in the poorer countries. But there were few nongovernmental groups that exclusively sought international solutions for population issues. With this in mind, Fornos approached Shaw, then with the Population Institute. Fornos recalls, "I gave him the idea that maybe we should establish a citizen action network called the Population Action Council. Rather than incorporate a new entity, would the Population Institute create a new division that would focus exclusively on international population issues? The board of directors voted yes. I would be the head of it and would have my own board of directors who would be advisory to the institute's board. I would have to raise all of the money to fund activities and staff. The Population Institute's board reserved the right, if it got too controversial, to separate us from the Institute."

Fornos and his Population Action Council began operations in June 1978, with a staff of three, including Fornos. In 1979 they played a seminal role in convincing the U.S. Congress to support family plan-

ning in Africa and, in 1980, were among the key players who kept the
Office of Management and Budget from eliminating the next year's
funding for international family planning projects. At the 1980 world
conference of the United Nations Decade for Women, the Population
Action Council entered a politically charged atmosphere that was the
result of disputes over the plight of Palestinian women and condem-
nations of Zionism. The council guided the delegates to consensus on
a declaration, not narrowly focused on the Middle East, but broadly
advocating reproductive freedom for all women.

These and his other accomplishments, as well as the overwhelmingly clear need for nongovernmental organizations in the international arena, enabled Fornos in 1980 to convince the Population
Institute's board of directors to adopt his recommendation that the
institute eliminate its domestic activities and focus entirely on international population issues. The institute's sole responsibility would be
to create awareness and marshal the resources to respond to requests
from third world nations that sought population assistance; it would
do this by building a domestic constituency in the United States that
would advocate the allocation of those resources. In May 1982, Rodney
Shaw retired and Fornos was appointed president of the Population
Institute.

I have known Werner Fornos since the early 1980s, but I didn't visit the
Population Institute until March 2001. As he and I walked out of the
Fornos residence several blocks from the institute's office in Washington, D.C., Fornos pointed to his house number—218—and said, "My
address is a constant reminder of what it takes to win" (218 votes is a
majority in the United States House of Representatives). A brisk five-
minute walk brought us to the Population Institute headquarters, an
attractive three-story townhouse painted gray with blue trim, the col-
ors on the emblem patch of the Twenty-ninth Infantry Division.

By eight-thirty Monday morning we had checked in at the institute and walked to the cafeteria of the Supreme Court building to
meet with several activists including Jane Roberts, a Population

Institute community leader from California, who was in town to talk with legislators. Fornos had courted Jerry Lewis, a Republican representative from San Bernardino, California, for the better part of two decades before Lewis agreed to never vote against any legislation that supported family planning and women's health. The situation had changed with the George W. Bush administration in power, and Lewis was being pressured to vote against international family-planning assistance. As a member of the Appropriations Committee, he was a key player, so Jane Roberts and her colleagues had come to let him know his constituents supported funding for population programs.

As Representative Lewis's wavering position indicated, things had taken an unfortunate turn on January 22, 2001, for those seeking to implement the Rockefeller Commission's recommendations that would reduce unintended pregnancies by providing couples with the information to help them decide when to have children and how many to have and the means to follow through on these decisions. On this day, his first Monday in office, Bush reinstated the so-called Mexico City policy established by President Reagan in 1984 and subsequently rescinded by President Clinton in 1993. Also known as the gag rule, this policy requires "nongovernmental organizations to agree as a condition of their receipt of Federal funds that such organizations would neither perform nor actively promote abortion as a method of family planning in other nations."

Werner Fornos and a host of other people and organizations have vigorously pointed out fatal fallacies in the gag rule that make it impossible for this policy to reduce the number of abortions while still promoting family planning, which were Bush's formally stated goals for reinstating it. Abortion is not a method of family planning but rather a procedure employed primarily because family planning failed. When family planning fails, there are two causes: contraception is not practiced, or the contraceptive method employed has not prevented pregnancy. When funds are denied to organizations that already cannot meet the family planning and health needs of people in developing countries, several interrelated things tend to happen. Fewer couples acquire the knowledge and the means to have the number of children they desire. The result is more unintended pregnancies, which in turn lead to an increase in both abortions and family size. More abortions, especially when performed illegally in makeshift facilities, mean more women will suffer and die. Increased family size and the resultant larger population will mean each person will have a smaller share of the resources available, often in a situation where resources are already limited. The environment is put under more stress and life-support capacity is further eroded. Poverty increases, which generally reduces the quality of life for the poor who experience it. Thus, reinstatement of the gag rule is bad for the environment and for people. As Fornos concluded decades ago, reproductive issues are cluttered with misunderstandings—for example, all women have children because they want them—and inaccurate information—for instance, education about family planning increases pregnancies—that impede the resolution of other pressing problems such as preventing famine or providing meaningful economic opportunities for all people.

In my assessment, this situation persists because cultural perceptions often change slowly or not at all, even with new information. Although the natural sciences have elucidated the answers to fundamental questions about human life and our relations with other organisms, many people cling to long-held beliefs and ideas that contradict or have nothing to do with empirical knowledge. As a consequence of holding such beliefs and ideas, people often ask the wrong question and the answer may not have a relevant context. For example, in the abortion debate, the first question some ask is: "When does human life begin?" Since life is a continuum, the answer—when the human species originated—is unrelated to abortion. Moreover, in the near term the question is biologically nonsensical because all human cells are human life and contain the information to make a human being. However, in a discussion about a specific situation relating to abortion, the first question could be: "What are the consequences for the fetus, mother, father, and relatives, as well as for the larger society, of conceiving this embryo and then allowing it to develop?" No single approach to human fecundity will fit all individuals, situations, or cultures. A humane society will grapple with the complexity of the issues and provide individuals and society with the greatest degree of freedom possible. In light of the carnage and havoc humans are inflicting on the biosphere, its creatures, and themselves, the most pressing and relevant question may be: How do we arrive humanely at a population size and a set of lifestyles that maintain life-support capacity, not only for humans, but for all life?

I had come to Washington to experience firsthand one of the Population Institute's unique programs: the Educate America Campaign. This campaign started in 1983 with the objective of visiting seven states every year. Each visit was to be several days to a week and include at least twenty-one events. As Fornos related, "The ideal schedule is an early morning radio show, a TV interview before 7 AM so that it could be included in the national news, a morning university talk, a lunch talk, a newspaper editorial meeting, an afternoon university talk, and a five o'clock news interview." For a number of years Fornos did the events solo after they had been set up and test run by a staff assistant. Now, he and Moyne Gross, an original staff member of the Population Action Council, do them together, but staff assistants still prepare these events.

After breakfast at the Supreme Court building we returned to the Population Institute for a staff meeting and a final briefing for the Educate Virginia road trip. At eleven o'clock Fornos had a live-radio phone interview with a station in Harrisonburg, Virginia, to kick off the Virginia segment of the Educate America program. We left the Population Institute for lunch at the National Press Club, where Fornos was to give the keynote address to several hundred teachers attending a Close Up Foundation conference. Close Up is a not-for-profit, nonpartisan citizen-education organization that helps people get involved in government, learn about the political process, and understand national and international issues. Each year about 25,000 educators and students come to Washington, D.C., to participate in its programs. This particular conference for teachers highlighted population issues. With the composure and skill of a master tight-rope walker, Fornos led the audience through the numbers and their meaning. Flashes of humor provided relief from the deadly serious message he was delivering: "If we do not change course, we will have environmental Armageddon."

Moyne Gross picked us up outside the National Press Club, and the three of us headed south to Richmond. As the congestion of the capital receded, Fornos handed me a three-ring binder, saying, "Here is the detailed itinerary and support materials. Look them over." The binder weighed several pounds and was perhaps three inches thick. The briefing manual provided, for each event, comparative population and environment statistics, directions, contacts, size of audience, venue, press releases, op-eds, and particular things to note or do. It served two functions: to prepare Fornos for each event and to educate the staff assistants who are part of the institute's Future Leaders of the World program.

From its beginning, the Population Institute has counted on young people to help plan, design, and implement its programs. In 1974 it sent forty journalism students to serve as reporters at the World Population Conference in Bucharest, Romania. Fornos told me that the combined airtime and printed material from these youth reporters exceeded that of any chain of wire services at the conference. The first young people to work at the Population Institute in the late 1970s and early 1980s searched state laws to identify obstacles to women's reproductive health, including denial of basic rights and access to contraceptives. The research compiled by these young staff members ultimately led to changes in a number of antiquated laws.

A permanent internship program of six new staff assistants every six months was initiated in 1980. These interns were instrumental in setting up the institute's We Care network of volunteer state and congressional district coordinators, many of whom are professionals—

doctors, educators, and religious leaders—with their own professional networks. With the initiation of the Educate America Campaign, using the existing We Care networks, interns went out into the states to make additional contacts at colleges and universities, as well as with numerous service organizations that wanted to sponsor talks on population issues. Through these projects and others, the interns learned about both population issues and how to conduct an information and legislative campaign, while at the same time substantially expanding the Population Institute's reach. Staff assistants, now called fellows, serve many roles, including media coordinator, public policy assistant, field and conference coordinators, and World Population Awareness Week coordinator. A few hundred applications for these positions come in from around the world each year, and Fornos interviews about forty applicants, selecting six for year-long fellowships.

Almost coincident with the start of the new intern program, Hal Burdett joined the institute staff in 1979 as director of information, and the bimonthly *POPLINE* was initiated, with Burdett as executive editor. It quickly became the largest circulation publication in the world dealing exclusively with international population and development issues. Today *POPLINE* provides evaluations of facts and public policies concerning world population issues to more than 2,000 newspapers worldwide, as well as to U.S. policy makers, correspondents in 174 countries, and other interested parties.

By seven-thirty the next morning, Fornos was greeting members of the Byrd International Airport Rotary Club. About thirty Rotarians ranging in age from 30 to over 90 got the same message as the teachers did the day before but uniquely packaged for Richmond and the business community. Fornos believes that couples have the right to use whatever means of contraception they choose, and, this morning, as in all of his talks, he clearly stated his agenda to "empower couples with contraceptive knowledge from natural family planning for those for whom it works, or abstinence for those for whom it works, to modern medically approved methods of family planning." In response to this statement a 91-year-old Rotarian's face broke into a smile that caught Fornos's attention. After the event Fornos asked him, "Why the big smile about halfway through the talk?" Smiling again, the man replied, "Natural family planning only works for sexually inactive couples over sixty-five." Fornos, clearly enjoying the humor, replied, "Well, whatever works."

The next stop on the tour was a geography class at James Madison University in Harrisonburg, Virginia. The core elements of his talk were the same, but Fornos tailored them to the youthful audience. After the stage had been set, Fornos asked, "How many of you are under twenty-four?" Almost everyone raised a hand. He continued: "The reason I ask you that is because you are in a very unusual age group. Never before in the history of the world have there been 1 billion 15- to 24-year-olds. You are about to enter your reproductive years. How well you are able to exercise the awesome responsibility of parenthood, how you determine the number of children you can love and take care of, and how you control your own fertility are not matters of governmental control. Rather, they depend upon having the opportunity to make choices. It will mean the difference between our having a better quality of life for our children and our grandchildren, or environmental Armageddon. It is that simple. I am confident we are going to make the necessary course corrections. I have absolute faith that your generation is going to do a better job at managing this planet than we have done. This lecture has not been about people as bald as I am but about your future and your children's future. I think you have a stake in that."

Fornos then held up a stack of volunteer cards for the students to fill out, saying, "What I am asking you to do for the sake of the planet is to volunteer ten hours per year." He then suggested many possible courses of action: write a letter to the editor, call an elected official, educate others, do anything that will show that you care how we leave the planet for the next generations. This is Fornos's mission—to mobilize as many people as possible, especially youths, to get actively involved in stabilizing the human population.

Like a Swiss watch, Fornos's week of scheduled events ran smoothly, without a hitch, one after the other—Buchanan Rotary Club dinner talk, back to James Madison University for another geography class, WHSV-TV interview, editorial board meeting with *Daily News-Record*, biology department seminar at Eastern Mennonite University, editorial board meeting with the *News and Advance*, talk at Sweet Briar College, and others. Some audiences were friendly, others not so friendly; however, Fornos's message and his responses to questions are always to the point: "We can stabilize the population at eight billion if we provide the services. . . . This is the only humane option we have." When asked why many in the United States do not appear to accept the importance of global population issues, he wastes no words: "We coddle the comfortable and ignore the afflicted." His response to questions about the Christian right was often equally terse: "The Christian right? They are neither right, nor Christian."

Fornos is perhaps at his best among those who seem truly to believe that couples, and women in particular, do not have the right to regulate their fertility. A private meeting with the managing editor of a city newspaper is typical of such an exchange. The newspaper editor was in awe of the late Julian Simon, a professor in the business school at the University of Maryland. The newspaperman agreed with Simon's arguments that the size of the human population has no foreseeable limit, everything is getting better, and more people are resources, not problems. Furthermore, Paul Ehrlich's predictions as outlined in *The Population Bomb* have not come true, Simon claimed. It is not overpopulation that leads to starvation and poverty, but an economic system that results in these social ills. Family planning is a sinister form of population control, and its agenda is eugenics. The editor espoused all these opinions and was absolutely against the government's using taxpayers' money for the explicit purpose of giving women the means for contraception or for committing the "mortal sin" of abortion, since conception and embryo survival were God's choices to make.

Fornos listened attentively to this newspaperman's arguments, then countered with facts: "We now have eighty-six nations that cannot feed themselves, and twenty-three of them are free market economies. Countries like Bangladesh have populations with just too few resources for any system to deliver the good life." The editor then acknowledged that population is a problem when the economy cannot grow. Fornos went on, "The efforts of educators like Paul Ehrlich and of family planning programs like Planned Parenthood in this country and abroad have been so successful that they are the reason that many of the worst predictions have not occurred more widely." He then cited cases of famine, civil strife, ethnic conflicts, and impoverished natural resources in Africa and elsewhere to establish that Ehrlich and others were sadly all too right. The conversation was never unruly and Fornos did not respond to the other man's outlandish and inflammatory jabs. Fornos closed his serious discussion by stating his fundamental belief: "It is a basic human right to decide on family size." But the editor had the last word, as he explained his anti–family planning position, "Your efforts are to stabilize population by giving people the means of voluntarily controlling their fertility. What if it doesn't work, and we have resource depletion. Then you in the population control movement will advocate coercive control of fertility. This is the logical flaw in your program." It was his own logic, however, that defied explanation: to state that the number of people is not the problem and then to conclude resource depletion can be caused by a growing population is not logically consistent. Furthermore, Fornos never advocated coercive measures but in fact supported the opposite. As Fornos and I walked to the car, I marveled at the brain's ability to rationalize any conclusion, if belief is strong enough. The editor believed that women did not have the right to regulate their fertility and therefore distorted data and "logic" to conform to this belief.

In 1983 the Population Institute declared the last week in October to be World Population Awareness Week. Each year libraries, high schools, colleges, and other institutions are provided materials and encouraged to focus on a particular population-related issue for the week; some of these have been: Saving Women's Lives in 2000, Population and the Urban Future in 2001, Population and the Next Generation: Youth and Adolescents in 2002, and Water: Our Most Precious Resource in 2003. About 80 countries participate in these programs, and almost 500 nongovernmental organizations—including the National Audubon Society, the League of Women Voters, the Christian Children's Fund, and the American Association of University Women—are cosponsors. During the five days of World Population Awareness Week, along with other media events Fornos typically gives talks in five states.

Long ago Fornos understood that if the media aren't covering an issue, people won't know about it. In 1980 the institute initiated the Global Media Awards to draw worldwide attention to population issues. Steadily gaining in visibility over the years, the award nominations peaked at about 1,000 when the world's attention focused on the World Population Conference in Cairo in 1994. In 2003 when no major population conferences were held, some 150 nominations were made. The awards ceremonies are both media events and educational. Fornos invites a dozen or more influential people to join the award recipients in a two-week tour of family planning and health care facilities in a developing country, which usually begins with the ceremony itself. Delegations went to Cuba in 2002 and Sri Lanka in 2003. In the category of best television documentaries, Turner Broadcasting has consistently won because of Ted Turner's personal commitment to produce documentaries on population, environment, and resource issues to be aired during prime time. Although the data on this have not been collected, it seems likely that the Population Institute through all of its programs is responsible for more media coverage of world population issues than any other single organization.

On our Virginia road trip Fornos spoke at four Rotary clubs, but his relationship with the organization is long standing. He has been an active Rotarian for decades and has nurtured the Rotarians' commitment to population issues. These efforts culminated in 1999 when

Rotary International's board of directors made growth and development issues a priority. In March 2001 Fornos spoke in Brasilia in central Brazil at the Third Rotary International Presidential Conference on Population and Development, as he had at the Rotarians' earlier two conferences.

In another international arena Fornos has taken a Population Institute delegation to every major United Nations conference on population and the environment since the Decade for Women conference in 1980. In recognition of his tireless devotion to population concerns, Fornos was honored as the individual United Nations Population Laureate in a ceremony at the United Nations headquarters in New York City on June 18, 2003.

During the week I traveled with him, Fornos told me about the Population Institute's newest initiative: the Village Contraceptive Empowerment Program. In 1999 the institute in conjunction with MEXFAM, a family planning association in Mexico, ran a pilot program to provide three impoverished communities with a one-time, one-year $6,000 grant for reproductive and empowerment activities. Its success, as Fornos relates, has led to "the most exciting program at the Population Institute. We are just learning to crawl with our villages program. We are scanning the earth to identify the one hundred poorest villages with the lowest literacy, the lowest employment opportunity, the lowest per capita income, the most inferior health system, and a high total fertility. As we identify these villages, we are looking into a new outreach program with America's family foundations, of which there are 460,000. We are encouraging individual family foundations, as a legacy, to make a one-time $6,000 contribution to improve the quality of life in one of these villages, working in partnership with quality nongovernmental organizations who are credible and effective in these various countries. It will be a shot in the arm to show communities that there is a better way into the future and that they can reach for that better way."

In his 1987 book—*Gaining People, Losing Ground*—Fornos debunks four demonstrably false concepts, ideas, or beliefs that are pervasive in the United States: the population explosion is over; population growth has no impact on economic development; free enterprise is the best contraceptive; and the real population problem is Western depopulation. The newspaperman mentioned earlier subscribed to all of these, and, judging by many of the actions taken by regional and federal governments in the United States over the past two decades, so do many others. I wondered how Fornos could be so positive about the future when the tremendous efforts of untold thousands had apparently failed to change so many minds about these myths. I asked him, "Your lectures are filled with a litany of devastatingly negative population and environmental facts. You have just told me that these four myths are very much alive and empowered with new vigor today. Yet, unlike many, you are overwhelmingly optimistic. Why?"

Fornos's response summarizes the extraordinary progress that human beings have made in slowing the growth of the world's population over the past twenty-five years: "Not everybody has been as involved as long as I have. You have to realize that when I got involved in population, less than 5 percent of couples in the world were consciously preventing unintended pregnancies. Today that has climbed to over 50 percent in the developing world and much higher in the developed world. The average family size in the developing world back in 1972 was six children. Today it is down to three. Nearly seventy countries have reached replacement level of fertility, or less. In the early 1990s we were adding almost 90 million people each year. Now we are below 80 million. I was in Colombia in 1972 at the presidential palace for the mother of the year award—she had twenty-four children. In 2000 the mother of the year had two children and both daughters had graduated from medical school. Eighty percent of Roman Catholics who have access to them practice and use modern family-planning methods.

"The obstacle to solving the population problem isn't demand or technology. We have the demand. Credible demographic health surveys establish that 350 million women want no more children—most did not intend to have their last pregnancy—or they want time between pregnancies, but they lack the education and the affordable means to do anything about it. We have the contraceptive technology, although I wish there was greater commitment to finding better male technology. Now we're sadly dependent only on condoms and vasectomies in regulating male fertility. But couples have enough choice to find the method that is best suited for them. What we lack is political will. It is through the mobilization of political will that we get sufficient funding to empower those couples in the world who want to regulate their fertility. If we can work together in a burden-sharing effort between the industrialized and the developing world to fulfill the demand for help, then we will achieve a population stabilization at eight billion. Heaven only knows where we will go if the mobilization of those resources isn't there. To me that is reality, not undue optimism. That is the engine that drives me to achieve population stabilization."

And drive him it does. The Educate America Campaign has been to thirty states three times and to the other twenty twice. This program alone, which takes about two months of Fornos's time each year, has required that he participate in over 2,000 events, more than a hundred each year, for the past twenty years.

On the last day of our week together, Werner Fornos and I arrived at the Population Institute in mid afternoon for a staff meeting and debriefing. As I walked up the stairs to the second floor, I began reading the Population Institute–awarded Legislator of the Month plaques that hung on the wall. I counted 31 Democrats and 17 Republicans among the recipients. Fornos is a consummate politician, but he is not partisan, just as population and other environmental issues are not partisan. During our travels he told me that his plan is to campaign in the last twenty states in the third round of the Educate America Campaign and then retire from the Population Institute. When he does retire, I don't expect he'll be "coddling the comfortable or ignoring the afflicted." Instead, I suspect he'll continue to work toward realizing his fundamental belief that world population can be humanely stabilized.

SEVEN

Living in a Finite World

Herman Daly and Economics

> Economists overwhelmingly agree that (1) economic growth, as
> measured by GNP, is a very good thing, and (2) that global eco-
> nomic integration via free trade is unarguable because it contributes
> to competition, cheaper products, world peace, and especially to
> growth in GNP. Policies based on these two conceptually immacu-
> late—and interrelated—tenets of economic orthodoxy are reducing
> the capacity of the earth to support life, thereby literally killing the
> world.
>
> HERMAN DALY, *Beyond Growth*

President Harry S. Truman was frustrated by the advice his council of
economic advisors was giving him. On the one hand, its members
would tell him, you should raise interest rates. On the other hand,
they said, you should lower them. In exasperation Truman pronounced, "I'm tired of this one hand, other hand business. What I
want is a good one-armed economist." In response to this story, told
at a public forum by another panel member, the economist Herman
Daly leaned forward in his chair and, with a slight forward motion of
his upper body, he propelled his right arm up—his only arm. Placing
his hand on the back of his neck, Daly commented, "Well, I was too
young for Harry Truman, but I am here now. His prophecy has been
fulfilled."

Why are economists equivocal? After all, even in Truman's time,
economics was paramount among the social sciences. The discipline
achieved its academic and political status in large part because, in an
attempt to be a science like physics, it grounded itself in rigorous
mathematics to describe market and human economic behavior. In
the eighteenth and nineteenth centuries—when Adam Smith, David

Ricardo, Thomas Malthus, and other early economists were laying the foundation for economics—mathematics did not dominate economic theory. It was Francis Edgeworth in 1881 with *Mathematical Psychics* who tipped the balance. To apply mathematics, Edgeworth simplified the world with an assumption that allowed human behavior to be quantified: "every [hu]man is a pleasure machine." Edgeworth's proposal that economics could quantify not only physical things but also economic behavior was attractive. For more than a century, mainstream economists have embraced mathematics for the credibility it gave their endeavor. In emulating physics, they sought fundamental laws for economic behavior that would allow scientifically based prediction. Unfortunately, economists trusted their mathematically rigorous models without authentic validation in the human world. As a result the sound decision to employ mathematics in the discipline has led to some huge unforeseen problems as economics gained stature and the market economy became widely accepted as the central organizing principle in industrial societies.

THE "SCIENCE" OF ECONOMICS

Physics is a predictive science because the operation of fundamental laws like that of gravity and those that describe the behavior of molecules and particles are based on mathematical models and analyses of direct observation of uniform physical entities. Within a class of fundamental entities like electrons, neutrons, and protons, and the chemical elements they aggregate into—hydrogen, helium, oxygen, calcium, and the rest—all members are identical. Once the behavior of the hydrogen molecule, or any other molecule, has been observed and recorded, the behavior of all hydrogen molecules can be predicted. It is the sameness of the members of each entity and of their physical interactions that enable physicists to predict future outcomes in certain situations with a high degree of certainty.

Individuals and the organizations into which they aggregate are the fundamental entities, or actors, in economics. They, however, never behave with the certainty or the predictability of those in physics, because unlike physics, no universal, fundamental laws of behavior for economic actors have been found. And I suspect none will be. The basis for this failure is the complexity of the elements in any economic system and the adaptive interactions among them. Take, for example, the powerful impact of human desires and emotions on the market. No two humans exhibit the same behavior, and human behavior as expressed within a community is strongly influenced by cultural and environmental factors. Hence, there is no way to predict just exactly how a group of humans will respond to an economic situation within a culture or, especially, across cultures. Despite this reality, economists have created and refined over the years a caricature called *Homo economicus,* or "economic man." To this hypothetical human they have given universal traits. Economic man is entirely self-centered, has insatiable wants, seeks to satisfy his own wants only through consumption, and is considered "rational" because he conforms consistently to these traits.

This hypothetical person is the fundamental element in economic models used to predict individual and aggregate economic behavior. Economic man behaves to maximize his happiness, or what economists call utility. Happiness is not easily quantified, but money is. So, for convenience of analysis, utility became equated with money. Most economists accept that people have preferences that do not conform with those of economic man and that beyond some base level of money (resources) more money does not increase happiness. Nevertheless economic models have yet to accommodate these facts, and in practice money is assumed the universal measure of utility.

To estimate a country's economic health, a government tracks its Gross Domestic Product (GDP: the market value of all final goods and services within a country over a given period). Other measures of utility and economic health have been suggested by economists and others—Measure of Economic Welfare in 1972, Economic Aspects of Welfare in 1981, Index of Sustainable Economic Welfare in 1989, Living Planet Index in 2000—but these alternative measures of well-being have made little headway in replacing GDP. They are messy, inconvenient, and more difficult to assess than the dollar value of market transactions, especially the latter two that attempt to assess sustainability and include aspects of life support.

Economics has inconsistencies in its theory and practice and in the behavior of actors in economic systems, because the discipline has not dealt with the actual nature of its fundamental elements. Each individual is truly unique and exhibits to varying degrees the character traits attributed to *Homo economicus* but also traits such as cooperation, self-sacrifice, sensitivity to others, and generosity, which are not ascribed to *Homo economicus*. All of this human diversity makes universal laws of economic behavior and precise predictability highly unlikely.

Herman Daly, among others, criticizes the economic man abstraction because it does not describe human economic interactions found in a culture and, perhaps more important, across diverse cultures. In addition, the concept of economic man fails to reflect the truth that individuals rarely act in isolation but as members of several overlapping communities and as members of a larger society. People in a community interact in dynamic, ever-changing patterns that depend upon their and their culture's past, as well as their innate connection with each other and emotions with regard to each other. This means that economics in practice is unlike physics in that economics lacks uniform entities that behave according to fixed laws because every person is different from every other person and every aggregate of individuals is unique in composition, history, and its cultural and natural environments. Without reconciling the behavior of the actors in their models with the actual behavior of people and groups of people, economics cannot be a science.

Behavioral and welfare economists have begun to broaden the scope of their discourse to include anthropology, sociology, psychology, politics, and biology in order to create a common model of human behavior. When a unified understanding of behavior—based on empirical data acquired through repeatable experiments with actual people across cultures—is achieved, economics may become a science. Even then, although general characteristics and behaviors consistent with human nature may permit some general forecasting, I suspect precise predictions of future outcomes in such a complex adaptive system will still be unlikely.

Modern economics also fails as a science for another reason: contrary to modern economic theory and practice, the economy is constrained by the biosphere. The great classical economists—Adam Smith, David Ricardo, Thomas Malthus, John Stuart Mill—lived at a time when it was obvious that the natural world, represented by land, was as important to production as labor and capital—the equipment and structures employed to produce goods and provide services. In the twentieth century, economists lumped together traditional capital and "land," or natural capital—forests, fisheries, soil fertility, capacity to assimilate wastes, water, and other natural resources. This arrangement fostered in theory and in practice an equivalence of natural capital and human-made capital, a major tenet of neoclassical economics. The flawed logic of this capital equivalence permitted economists to assume falsely that the human economy subsumed the natural world merely as a substitutable factor of production. For the economist, the natural world became part of the total capital stock available for production, and its "products and services," as priced in market transactions, became its value. Nature was reduced to a commodity.

This assumption that all of the non-human environment was just part of the human economy resonated with the human sense of self-importance and the cultural belief that the natural world was intended for human use alone. Cultural acceptance and unrestrained economic growth made it easy for economists to accept unlimited economic growth as possible because the modern economy had not been constrained by the natural world. Belief in perpetual growth was tenable with a small human economy on a vast, unused planet. But, in the twentieth century, as the world filled with humans and their activities, the assumption of unlimited growth proved in conflict with the laws of nature. Herman Daly was among a handful of maverick economists who identified this conflict. Realigning economics with biological and physical reality became Daly's life's work.

ECONOMICS ON EARTH, A FINITE PLANET

Herman Daly did not always have one arm. When he was seven, polio destroyed the deltoid muscle in his right shoulder and made his left arm nonfunctional. Daly recalled, "I struggled for seven years to make that atrophied arm work. The nerves were just dead and to move that arm was like trying to wiggle your teeth. Finally it dawned on me to prune the tree and put energies where there was a better chance of success." His parents concurred and at fourteen he became one armed. Daly considers the shoulder as a bit of a handicap but having "just one arm was no big deal. I played tennis on the high school team and really enjoyed swimming. We lived for a while on the Texas gulf coast. I'd go swimming every morning and fish in a little skiff or off the pier. I think it was through the sea that the natural world reached in and grabbed me."

Neither of his parents had much formal education, but they encouraged both Daly and his sister (now also a professor) to go to college and beyond. But his father kept running into highly educated people who were not terribly capable in every-day life. Daly remembered with a smile, "He sometimes thought Ph.D. meant 'phenomenally dumb.' He would tell stories about Ph.D. students he hired in the family hardware store who didn't have sense enough to 'pour piss out of a boot, if the instructions were written on the heel' as he put it. So he had this ambiguous attitude about higher education and I guess I kind of picked that up from him."

As his parents didn't have much money, Daly was fortunate to be accepted to Rice University, which charged no tuition in the 1950s. He liked both the natural sciences and the humanities but couldn't decide which to study. He reflected, "I felt somehow the social sciences would be a good compromise, because I believed at the time that they had one foot in the natural sciences and the other foot in the humanities.

Economics would be the most useful since I had to deal with making a living, so I majored in it. Of course I was completely wrong about the social sciences having one foot in each camp. They had both feet in the air. Over a long period I discovered that was what was really wrong with economics. It had to get one foot into the natural sciences. And it needed another foot in ethics and humanities. My mistake became my life's project."

Daly didn't like school very much. In fact, by the time he graduated from Rice in 1960, he was tired of it. Reflecting on a road trip he had taken to Mexico City upon graduation from high school that opened his eyes to equity and social justice issues, Daly speculated, "If there had been a Peace Corps when I graduated from Rice, I would have joined." A summer spent doing paperwork at Tennessee Gas and Transmission made him know that "I just don't want to do that every day." So, for want of a better choice, he went to graduate school to study economics, a year at Rice, and then to Vanderbilt University for the Ph.D.

The neoclassical economics he learned made him believe that he could use it to make everyone rich. He then went to Uruguay to do his dissertation on that country's trade policy. While completing his Ph.D. dissertation, he was an assistant professor at Louisiana State University (LSU); this position was followed by a job with the Ford Foundation, teaching economics in northeast Brazil. The poverty that Daly encountered in Uruguay and Brazil was oppressive, but what really caught his attention was population growth. It was 1967, and Daly was living in the midst of the population explosion—3 billion then to 6 billion now—about which Paul Ehrlich would soon warn the world. It wasn't the burgeoning numbers, however, that most struck Daly, but the disparity between classes. The rich had few children whereas the poor had many. Daly saw that if the number of workers doubled or tripled every generation, thereby allowing the number of laborers to exceed available jobs, the laboring poor would never command decent wages. He would not be able to make everyone rich.

Through his concerns about population, Daly began to think

about resource use and depletion. It was during this formative time that he read Rachel Carson's *Silent Spring*. Suddenly, everything came together, as if the dark forest he had been stumbling through was suddenly illuminated. Daly saw clearly the ecological context within which an economy operates. As a graduate student at Vanderbilt, he had studied under Nicholas Georgescu-Roegen, but at the time he had not understood the full import of his mentor's obsession with thermodynamics (relations between heat and energy) and its connection with economics. The ecological context Carson gave him brought to light the significance of Georgescu-Roegen's focus on energy (the capacity to do work) and entropy (a measure of disorder), and the role of the economy as an agent that increases entropy.

Georgescu-Roegen's insights hinged on the laws of thermodynamics. The first law of thermodynamics states that energy cannot be created or destroyed, but it can change form. This is good news because forms of energy like chemical, electrical, and mechanical energy are interconvertible, or mutually convertible. The second law states that when energy changes its form, some of the capacity to do work, or available energy, is lost. This is bad news because it means that when work is being done, energy changes form and the available energy decreases, while entropy, or disorder, increases. All living organisms create and maintain order, or a low-entropy state, by using energy from the sun: photosynthesis converts light energy into the chemical energy of sugar, which in turn provides energy for essentially all life forms. In the process of creating and maintaining order by using the energy in sugar, organisms reverse the process of photosynthesis and release carbon dioxide and water as "waste products," as well as heat energy that has a reduced capacity to do work. Without a constant source of more low-entropy matter such as sugar, an organism dies.

Georgescu-Roegen was among the first eminent economists to conceive the analogy between human economies and organisms: economies, like organisms, use the energy and order in low-entropy matter (resources like wood, grain, meat, oil, coal, and phosphate fertilizer [concentrated phosphate]) to make or do things and in the process produce higher-entropy matter (wastes like carbon dioxide, water, and diffuse phosphate that is less ordered than concentrated phosphate). Following this analogy, he inferred that, in the same way that the laws of physics put limits on the biophysical world, these laws put limits on the economy: the economy, like an organism, cannot operate without the continuous use of energy to maintain order.

Georgescu-Roegen therefore argued that the rate at which matter and energy can flow through the human economy, or what is called "throughput," is naturally limited. Although he is correct that Earth does not have unlimited useful energy, one can contend that throughput is not limited in practice because the amount of energy available at any given time from nuclear sources, fossil fuels (stored solar energy), and sunlight itself is so immense that human economies cannot possibly use it all. This contention, however, may be irrelevant, because throughput is most likely constrained by biological limits, not availability of energy. Consider, for example, fossil fuel energy. Throughput could be increased 100 times or more by using fossil fuels at a much higher rate. This would substantially accelerate the increase of the heat-trapping gas carbon dioxide in the atmosphere, thereby warming the planet many degrees and faster than it is now being warmed. The resulting climate change would cause the extinction of many species and ecosystems, leading to the impoverishment of elements of life support and making present agricultural yields hard to maintain. The consequent die-off of many humans would severely stress global civilization, which would most likely collapse, yet the energy supply would not be depleted, just reduced a bit.

Earth and its biosphere fueled by energy from the sun constitute a finite system within which human activities occur. That is, the human economy is a component of the larger biological enterprise. This economy is constrained by the same laws of nature that impose limits on life in general. Consider the common bacterium *Escherichia coli* that lives in our intestines. Under ideal conditions one bacterium could increase to 2.2×10^{43} bacteria in two days with a mass of over 100 times that of Earth. To use another example, at the current growth rate (births minus deaths per unit time), human mass will equal the mass of Earth in about 2,000 years. This biotic potential is never realized, however, because living and nonliving factors in the environment curtail growth. Infinite growth is just not possible for any part of a larger, but finite, system. The universe may expand forever into the void, but Earth-bound entities will not.

When I began working with economists over a decade ago, my natural science perspective made it impossible for me to accept that economics should be taught as though continual growth were possible. However, modern economists have traditionally held a very different vision of reality from those who study how things work in the natural world, as illustrated by the following story.

In 1992 the World Bank published a report titled *Development and the Environment*. At the time Herman Daly was an economist at the World Bank, and, although not an author of the report, he was given opportunities to comment on drafts. In an early draft, a figure labeled "The Relationship between the Economy and the Environment" contained a box titled "economy" with an arrow labeled "inputs" going into the box and another arrow exiting the box labeled "outputs." Daly wondered, where was the environment from which the inputs came and into which the outputs went? He suggested to the authors that a second box be drawn around the economy box and labeled "environment." In a later draft the authors had drawn a box around the economy box, but the enclosing box was not identified in the figure or in the caption. Daly's comment on this revised figure was: "[T]he larger box has to be labeled 'environment' or else it [is] merely decorative, and the text [has] to explain that the economy is related to the environment as a subsystem within the larger ecosystem and is dependent on it." The final report had no diagram. Several months later, Daly had an opportunity to publicly ask Lawrence H. Summers, then chief economist at the World Bank, about a similar diagram in another institution's publication that was labeled as Daly had suggested. Summer responded, "That's not the right way to look at it."

Daly contends that the World Bank's approach to acknowledged social and environmental problems was, and continues to be, to ignore their direct links to the constraints of a finite world and to recommend more economic growth. Standard economic theory and practice persist in advocating growth, often disingenuously called "sustainable development."

Despite biological, historical, and theoretical evidence to the contrary, in their practice of economics most economists do not accept that the growth of the human economy is constrained because it is a subsystem of the larger biosphere that imposes limits on growth. Daly explained their view in this way: "The paradigm of a natural scientist is not the economist's paradigm. In contrast to the scientific view, they believe that the ecosystem is a subsystem of the economic system. Economists believe what is going on is not entirely a physical thing. The focus is on gross domestic product which is partly physical things and partly things like want-satisfaction and value that are not really reducible to physical dimensions. 'Look,' says the economist, 'all that the physical world gives us is atoms and molecules. We, with our knowledge, arrange all of these things in different ways. Atoms and molecules are nondestructible. We just bring technology to bear on this physical material of the world. We make of it whatever we want. If nature gives us a little limit, then we just use technology to substitute around the limitation. It is the technologically driven human system which is important and the natural system is drawn upon for raw materials.' It's crazy, but they believe it. It gets into the power of technology and knowledge that are believed capable of eliminating limits. Yes, the world is a finite system. Economists don't deny that. But we can do infinite things with it given the infinite power of knowledge and technology. We can make the economy grow around this seed or crystal of the natural world. We can turn material cycles faster. We can invent new things. The natural world is not a constraining envelope; it is a little seed from which we can grow into the void."

The difference between the two paradigms, in Daly's words, "could not be more fundamental, more elementary, or more irreconcilable."

Continual physical growth of the human economy is impossible because other components of the larger physical and biological system upon which the economy depends are consumed or otherwise compromised as the economy grows, thereby putting constraints on future growth. Neoclassical economists, such as Lawrence Summers, do not accept that the economy is a subsystem of the biosphere and do not acknowledge as true John Stuart Mill's 1857 assessment that an economy will achieve a stationary state rather than unlimited growth: economies will settle into a stationary state where population and human activities maintain an equilibrium with the rest of the natural world while improvements in technology and ethics persist.

For an economist to reject the concept of infinite economic growth means that the century-long belief in economic growth theory and practice has been wrong. To do away with the idea of limitless growth would render economics textbooks and much economic policy obsolete. For practitioners of a discipline to admit that such a basic idea is a mistake is perhaps too much to expect. Nevertheless, elements of modern economic theory—parts of microeconomics and market theory in particular—are useful in working toward what Mill called a stationary state and what Herman Daly coined the "steady-state" economy, a dynamic state of persistent innovation and change without the perpetual physical growth of human activities.

OPPORTUNITY COSTS IN THE MACROECONOMY

Microeconomics considers component parts—firms, workers, products, materials, wages, households—of the macroeconomy and deals primarily with one issue: efficient allocation of resources. If the cost to produce an item exceeds its price, the resources to make the item (labor, transportation, raw materials, ecological impoverishment, and so on) have not been efficiently allocated. This can be easily demonstrated by means of an example such as the production of shirts. For instance, if it costs $5 to make a shirt and that shirt can be sold for only $5, no more shirts should be made; if that shirt can be sold for only $4.50, too many shirts have already been made. Clearly, to produce a shirt that costs $5 to make but sells for $4.50 is not an efficient use of resources.

Market mechanisms establish the efficient allocation of a scarce item such as a shirt, a cup of coffee, transportation, and labor (price paid for time). Marginal analysis identifies fruitless economic activity: when the market value of the next unit equals the costs to make it, it's time to stop. Each microeconomic activity draws resources away from other activities: the $5 of resources used to make a shirt that sells for $4.50 wastes $.50 of resources that could have been used more efficiently for other things. That is, when you operate in a finite system, you choose among alternatives. These choices incur what economists call opportunity costs.

Opportunity cost is the key concept in microeconomics. Prices are supposed to measure opportunity costs, and opportunity costs presumably determine the choices people make. For each choice, the opportunity cost equals the most beneficial alternative sacrificed. Consider you have a limited amount of money that has to provide all your food for a week. You could have one very fancy meal, three fancy meals, or some food for seven days. A rational person would select food every day as the most beneficial. The opportunity cost of having food every day is not having one very fancy meal or three fancy meals, whichever you consider the next most beneficial choice. In the economist's model of human and market behavior, a rational person will never incur an opportunity cost greater than the value of the option chosen, nor will an efficient market ever lead to a choice that has such an opportunity cost. Opportunity costs exist because each microeconomic activity competes with other activities for limited resources within the total economy.

When Daly began his economic career, this fundamental concept of opportunity cost was not considered relevant to macroeconomics. At that time, the macroeconomy—the sum of all microeconomic activity—was not considered part of a larger biophysical system; in fact, it was viewed as the whole, not a part. In this view, if the macroeconomy is not bounded by the biosphere and can grow into the void, unlimited growth—as advocated by the World Bank, International Monetary Fund (IMF), World Trade Organization (WTO), and economists in general—is possible and "sustainable," and growth has no opportunity costs.

In the late 1960s Daly realized that the natural sciences, along with everyday observations, provided unequivocal evidence that the human economy is indeed part of the larger biotic enterprise, which is itself a subset of the even larger earth system. He has argued for more than three decades that such things as Earth's carrying capacity, pollution of the food web and degradation of topsoil, species extinction, ecosystem loss, and climate change establish that microeconomic-type analyses in the macroeconomy are essential to avoid huge problems in the future.

Daly sums it up this way: "Ecological limits are rapidly converting 'economic growth' into 'uneconomical growth'—i.e., throughput growth that increases costs by more than it increases benefits, thus making us poorer not richer. . . . As the macroeconomy grows in its physical dimensions (throughput), it does not grow into the infinite void. It grows into and encroaches upon the finite ecosystem, thereby incurring an opportunity cost of preempted natural capital and services. These opportunity costs (depletion, pollution, sacrificed ecosystems) can be, and often are, worth more than the extra production benefits of the throughput growth that caused them." Daly labels this "negative economic growth," because throughput grows while overall human benefits decrease, especially because of loss of life support.

Take, for example, the global fishing industry in recent years. *The Inexhaustible Sea,* a book representing conventional wisdom when published in 1961, purports that the oceans are an endless source of fish. We believed that this was true and acted on this belief. Today a million fishing vessels—an increase of 100 percent in twenty-five years—employing ever-more efficient techniques, are struggling to harvest fewer and fewer, and smaller and smaller, fish from the world's oceans. The anchovy fishery off Peru collapsed in 1973 after record catches in the late 1960s. The North Atlantic cod fishery is exhausted and mostly closed. White abalone (*Haliotis sorenseni*) in Mexican and California waters, which once had as many as 4,000 individuals per acre, is now near extinction with densities of 0 or 1 per acre. Throughout the world's oceans, populations of predatory fish— tuna, billfish, swordfish, codfish, flatfish, skates, sharks, rays—have been reduced by about 90 percent since the beginning of industrial fishing in the 1960s, and some species are teetering on the brink of extinction.

Oceanic ecosystems are being further impoverished because the techniques that efficiently harvest the commercial species also catch many unwanted organisms. In the shrimp fishery only about 15 percent of the catch is shrimp; the rest is "bycatch" and discarded, often dead. This bycatch is estimated to be 60 billion pounds of wasted squid, octopuses, turtles, rays, sharks, sea anemones, starfishes, and other sea life. Industrial fishing has collapsed or put in decline 13 of the 17 major ocean fisheries and pushed the other 4 close to or beyond a sustainable yield. More boats are chasing fewer and smaller fish at greater cost per fish, so that this global-scale economic activity is, in fact, Daly's negative economic growth. Moreover, industrial-scale commercial fishing is compromising the oceans' capacity for supporting life and thereby collapsing major oceanic ecosystems.

The longer negative economic growth persists, the more fragile life support, or Earth's carrying capacity, becomes. Thus, Daly's agenda is to move as quickly as possible to a dynamic and creative, steady-state economy. To do this, negative economic growth needs to be eliminated by applying microeconomic-type analyses to the macroeconomy so as to identify and eliminate those economic activities that cost more than the benefits they provide. Although this cannot be done with great precision, we can be broadly accurate. That is, our analyses of resource use through tools such as ecological footprinting—along with our knowledge of trends in climate change, species loss, and pollution—tell us we have considerable negative economic growth. Simply put, we are using water from a rain barrel, and before the barrel is empty, we need to allow it to replenish.

CAN NATURAL CAPITAL BE APPROPRIATELY PRICED?

We know we must curtail our impact on the environment, but to what degree is uncertain. The ultimate goal is to keep Earth's natural resources—forests, soil fertility, fisheries, fresh water, wilderness areas, assimilation of wastes, and so on—intact so the planet can continue to provide life support for humans and the rest of the biotic enterprise. Although many economists believe that human-produced capital— machines, trucks, highways, factories, and all the rest—can substitute for natural capital, Daly points out, "What good is a sawmill without a forest, a fishing boat without populations of fish, a refinery without petroleum deposits, an irrigated farm without an aquifer or river?" That is, natural and human capital are, by and large, complements, not substitutes. Yes, we can do some things like foul our water and then purify it, but, as New York City discovered, this is costly. Preserving the Catskill watershed, a major source of New York City water, for $1–2 billion is a bargain compared to the $6 billion or more needed to build the filtration plant required to clean the water—and the ongoing annual costs of operating such a plant.

This New York City water-supply situation illustrates another economic shortcoming that Daly seeks to correct. Potable water from the Catskill watershed is delivered to city residents for a fee that mostly represents the "value added" to potable water. The "value added" approximates the costs of the infrastructure needed to bring the water from its source to residents. The potable water itself is, for the most part, free. This is how economists treat all natural resources; the only value is the value that humans add to "valueless," or unvalued, stuff. Consider what happens when water becomes polluted. Energy and other resources are used to make the polluted water potable again. As a result the cleaned-up water costs more because value is added by the human manipulations that made the polluted water potable. Cleaning up polluted water is inefficient and expensive because energy is required for purification, and as a result there is less useful energy available to do other work.

Daly contends that paying for the use of natural capital is economically sound for two major reasons. First, if a natural resource is scarce but free, people waste it and it becomes even more scarce. Second, such resources in the public domain can be priced and sold so that their capacity to support the economy is used efficiently, and the money generated can be employed for the most pressing social and environmental needs. For example, until 1971, the city of Troy, New York, provided residents with free water, and predictably water consumption kept growing. When a modest fee per unit of water used was imposed, demand dropped by over 30 percent. Since the residents of Troy owned the source of the water and since it was priced close to the cost of delivery, the residents could have priced the water higher than delivery cost. If the water had been priced higher, the increased income to the city might then have displaced other taxes or been used in other ways to improve human welfare. This is exactly what Daly advocates.

Believing that all natural resources can or should be priced is, however, a dangerous idea. By setting a price for a natural resource, we presume that we know its total value, which in many cases is certainly not true. If we do not know a resource's full value, it cannot be priced correctly. For example, what should be the price of the Catskill watershed? What value should we ascribe to its clean water supplied to New York City? Its agricultural output? Its recreational use? Its aesthetic value? Its forests and the carbon they sequester from the atmosphere or the flood control they provide? Or the still-unidentified life support it provides? And how shall we calculate the "correct" price for all of these "goods and services" and those that future generations will need but are still unknown? Clearly preservation of some natural resources will be enhanced through market mechanisms, but many will not. In fact, many natural resources will be harmed, diminished, or eliminated by market forces. The challenge is to identify and appropriately price those elements of natural resources that the market can efficiently allocate while creating mechanisms that preserve those aspects of the natural world that have little or no currently accepted value yet may have unrecognized present or future value.

APPROPRIATE SCALE FOR THE MACROECONOMY

In the second half of the twentieth century the dominant objective of economic policy has been efficiency. In a market-driven economy with an efficiency-first policy, the following happens: (1) price goes down, (2) more resource is available, (3) more resource is used, (4) more uses for a resource are found, and (5) throughput goes up. Daly uses analogy to show that we need a measure beyond efficiency to assure sustainability, with a cargo boat as an example: "This absolute optimal scale of load [for a boat] is recognized in the maritime institution as the Plimsoll line. When the watermark hits the Plimsoll line the boat is full, it has reached its safe *carrying capacity.* . . . The major task of environmental macroeconomics is to design an economic institution analogous to the Plimsoll mark—to keep the weight, the absolute scale, of the economy from sinking our biospheric ark." To establish a safe load for a boat is a relatively easy calculation based on a well-understood branch of physics, and we have learned also through observation and experience, unintentionally sinking boats by miscalculation or overloading them. Determining an economic "Plimsoll line" for our global economy will be more difficult, however, and will require objective economic and ecological measures to warn us of impending danger. Here, too, the histories of successful and failed cultures can be helpful.

The central question for mainstream economists: Is the current level of global economic activity sustainable? Unfortunately, current economic models cannot answer this question; nevertheless, when the global economy collapses, its limits will be known. Most economists are not worried, however, because, historically, substitutes for scarce resources have been found and technologies have circumvented apparent limitations.

The data, as we have presented throughout this book, show that the ark is taking on water; yet, among mainstream economists who establish policy, Daly's recommendations, such as reducing throughput or assessing opportunity costs in the macroeconomy, have been ignored. He believes time is running out, and while we have sufficient natural capital and social resilience remaining, we need to set limits on the rate at which we use that capital. Energy (fossil and nuclear fuels) may be the element most easily controlled, and to limit energy production and consumption would constrain much of the rest of the economy and thus the throughput, as thermodynamics predicts. By setting limits on production and consumption, we would have a frugality-first policy in which energy prices would go up, less energy would be used and for fewer things, and throughput would go down, while the market would efficiently allocate the energy allowed to it. Over years or decades, depending upon the severity of the limits placed on energy use, this policy would stimulate the shift to other energy sources such as wind, geothermal, and solar, thereby making it necessary in order to create and maintain a steady-state economy to limit the economic use of all forms of energy. Unfortunately, with pervasive and powerful interest groups now championing low-priced energy—especially in the United States where the policy keeps energy cost artificially low through massive subsidies—even the first step of limiting fossil fuel and nuclear energy use seems hard to imagine.

MONEY SUPPLY IN A STEADY-STATE ECONOMY

In his academic meandering in the late 1970s, Daly came across a book titled *Cartesian Economics,* whose subtitle, *The Bearing of Physical Science on State Stewardship*, identified exactly what he was trying to understand: what are the limits that physical laws place on economics? The 1922 book written by Frederick Soddy, a chemist who received the Nobel Prize for his work on atomic structure, was a gold mine. In the second half of his life, Soddy abandoned chemistry and turned to critiquing economics because he was convinced that defective economics would implement ruinous policies in the name of good science. Soddy foreshadowed Georgescu-Roegen in his assessment of the fundamental importance of thermodynamics to economics.

Soddy's critique of money was visionary. Money, he claimed, is one of four foundational elements without which modern civilization would not exist; fire, language, and the wheel are the other three. Money, however, is mysterious in ways that the others aren't—there is nothing equivalent to money in the natural world. The fundamental abstraction of modern economics permits us to create an artificial world in which everything has a value that corresponds to a quantity of a single unit—money. In this abstraction a dollar's worth of something is equivalent to a dollar's worth of anything else. This means time (labor), a piece of land, water, watching a movie, or the right to grow a genetically modified seed is fungible—its value can be expressed in monetary terms and it may be exchanged for money. This system yields extraordinary benefits. Money is symbolic as opposed to something of real value used in trade such as food or clothing. As a piece of paper or a number in a computer, money can be kept and then converted at any time into real things (wealth), through the act of purchase. Through this act, money represents purchasing capacity. Money does not spoil or wear out and its physical condition does not diminish its exchange value, although money can become worthless. Money is a huge convenience over barter; it is conveniently carried and transactions now occur electronically.

However, wealth exists, whereas money is an abstraction and logically disconnected, by and large, from the physical world, as Soddy pointed out. Take, for example, two pigs. They are wealth. They are physical entities. Yes, they do grow and more pigs can be generated, but their growth is limited by the flow of matter and energy, obeying the laws of thermodynamics. The situation is different with money. If we borrow money to buy the two pigs and have to pay interest on the debt, the interest accumulates according to a mathematical equation. The interest generated can grow to infinity, whereas the pigs cannot. Debt is governed by the permissive mathematics of compound interest, whereas real wealth is governed by the restrictive physical laws of thermodynamics.

We have been culturally conditioned to believe that debt and real wealth increase and decrease in the same way; yet, as Soddy's example of pigs versus interest shows, this belief is fallacious. We have an economic system whose unlimited growth of interest is an abstraction that has a poorly defined relationship to physical reality. The unlimited growth of interest creates quantities of money for those who collect the interest on debt, and they use this money to acquire real things. This growth of money, however, does not correspond to the limits placed on real things.

John Kenneth Galbraith noted, "The process by which banks create money is so simple that the mind is repelled." When a bank receives a deposit of money, the institution is permitted by banking regulations to put additional money in circulation based upon the reserve requirement. For example, if the reserve requirement is $10 for every $100 debt, the bank must keep in reserve only one-tenth of what it loans out. For a deposit of $100, the net result in the banking system is new loans of $900. Bingo, like magic, $900 is created. One thinks, "This can't be true. How can banks make money out of nothing?" The mind may be repelled by this process, but in fact it is mandated by human convention. This additional money is born as a debt to be repaid with interest. When the debt is repaid, the money supply declines by the amount loaned. Therefore new loans to finance new investments must continually be made to keep the money supply from shrinking. In other words, our debt-based money supply requires economic growth to avoid shrinking.

The relationship between the money supply and economic activity is complex, because the behavior of the economic actors depends upon myriad factors. Notwithstanding this complexity, the money supply and economic activity are often positively correlated: as the money supply increases, economic activity also increases, which in turn leads to the use of more physical resources. This is fine so long as the activity can be sustained by the flows and stocks of resources. In a simple example, from a fish stock or a forest stock that provides a variable annual growth averaging 25 units, it is possible to harvest something less than 25 units each year without compromising future harvests. If economic activity is fueled by annual consumption of more than 25 units, then the stock will be diminished and future harvests compromised. A growth economy is a wonderful thing for those present in the beginning when the size of the flows and stocks dwarf the scale of economic activity, which at the onset is more than sufficient for current users. The situation for later generations that have to survive when stocks and flows have been reduced, however, is a different matter.

The formidable challenge for economists is to effect, with manageable turmoil, a transition to an economy that is compatible with a level of resource use that does not impoverish life support and at the same time sustains people by providing for their needs. This is easier if the money supply is not interest-bearing debt and is not otherwise growing. Daly contends humanity has a choice: to continue advocating economic growth until we reduce life support to the point of civilization collapse, or to carefully transit to a steady-state economic system in which economic growth stops.

COMPARATIVE ADVANTAGE IS ABSOLUTE ADVANTAGE IN PRESENT-DAY GLOBAL TRADE

Herman Daly is used to being ignored for his contrarian views: "I taught for twenty-one years after I left Brazil. The first half of that time was good. Then I began to attract Ph.D. students who were interested in the kinds of things I was doing. I found much to my dismay that I really couldn't get a Ph.D. student through a faculty committee, because it considered the research very questionable. Certainly not good economics; really not economics at all in their opinion. After a couple of bad experiences, I became very upset with what I considered the unfairness, narrowness, and simply bad treatment my students had to endure. I just wanted to get out of academia. As I waited a few years for early retirement, an opportunity to work at the World Bank came along."

Robert Goodland, an economist at the World Bank, thought Daly could nurture the bank's nascent environmental interests by being a liaison between economists and the few ecologists at the bank. Daly retired from teaching at LSU and began working at the bank in 1988.

It was a huge adjustment. In academia you are your own boss and the goals are to be creative and to do it right. At the bank the important things are making a deadline and doing what your boss wants. Daly recalls, "It was good discipline for me and I met lots of wonderful people. After six years I realized that the World Bank did not merit the people it had working there. It was not going to face up to the problems of the world. If I stayed, it would be more of the same. I gave up on the World Bank. This institution is really never going to get it. They are totally devoted to growth. I hope I'm wrong."

Daly was fifty-five and his small retirement income couldn't pay for the basics. He applied for a faculty position in the University of Maryland's School of Public Affairs at the suggestion of Peter G. Brown who had founded the school. After interviews, the faculty was seriously split between Daly and another person. The differences were so contentious that no one was hired for the tenured position. Brown persisted, and eventually Daly was hired as a research associate with a ten-year appointment.

Several years passed and Brown decided to put Daly up for a regular tenured position. The faculty reviewed his case and declined to grant him tenure. Then in 1996 Daly was awarded Sweden's Honorary Right Livelihood Award (considered the Nobel Prize for environmental disciplines) and the Heineken Prize in Environmental Science for being one of the first economists to focus on environmental problems and for founding ecological economics, an emergent field that broadened the economic conversation to include the science of ecology. Daly recalls, "Peter Brown came to me and said, 'Well, you ought to try again.' So we put the request for tenure review in again and this time it went through." Although it seems that a research university would honor and readily grant tenure to a person whose work had been recognized by prestigious awards, Daly's tenure did not come without controversy.

Daly is quick to agree that changing one's mind comes hard, especially about a long-held position: "Unlearning is hard—getting rid of a cherished wrong idea is difficult at best, if not impossible. One of my main unlearning experiences was during the writing of the book *For the Common Good* with John Cobb [a theologian at the Claremont Graduate School]. Even after I had come to question economic growth, the *summum bonum* of economics, I still retained a commitment to free trade based on comparative advantage. I thought it was fundamentally well founded, and I was proud of this on behalf of all economists. In writing the book with John, each of us would draft a chapter and then exchange them to be criticized. He wrote the first draft on trade. When I got his draft, I was horrified to see his critique of free trade. I thought, 'Well, John just doesn't understand comparative advantage and it is up to me to explain it. He will see the light.' In fact, I was quite sure about that. But John is a very smart guy and I knew he wouldn't let me get away with anything. I started going over it very carefully. As I prepared myself to refute what he had written, I began to discover, 'Gee, that refutation isn't really right. Also John could have made his point even stronger by saying this instead of that.' Instead of discovering flaws, I found other reasons to support what he had written. I went back and read Ricardo again. Imagine that! [David Ricardo was the late-eighteenth- and early-nineteenth-century economist who established the fundamental principles of comparative advantage associated with international trade.]

"I was dumbfounded to see a fundamental assumption of Ricardo's for comparative advantage was that capital had to be immobile between countries. [Labor is also assumed to be immobile.] Where is that in the textbooks? Not there! I think you can find it in earlier textbooks in the distinction between international and interregional trade where interregional trade is governed by absolute advantage because capital is mobile within the region. Unregulated international trade, or free trade as we call it now, is governed by comparative advantage only if capital stays within countries. Free trade becomes economic integration if capital is mobile, because comparative advantage becomes absolute advantage. Different countries become like different regions of the same country. You lose the situation where both parties benefit. One party can get all of the benefits."

The profound significance of Daly's rediscovery of Ricardo's assumption that capital does not move among countries can be appreciated by a rudimentary explanation of comparative advantage versus absolute advantage in international markets. Assume that only goods, not labor or capital, move across borders and that all costs are represented by equal hourly wages. Assume also that countries A and B can make cars and computers and, in addition, that country A makes a car in fewer hours than it does a computer (each car takes 80 hours of labor to make and each computer takes 90 hours), whereas country B makes a car in more hours than it does a computer (each car takes 120 hours and each computer takes 100 hours). Under these conditions, country A benefits by trading cars to country B for computers, which also benefits country B because of comparative advantage. That is, country A has a comparative advantage only in cars, because its internal ratio of costs is lower for cars but not computers, when compared to country B's internal ratios of costs, as follows:

COUNTRY A   cars .88 [80/90]   computers 1.13 [90/80]
COUNTRY B   cars 1.20 [120/100]   computers .83 [100/120]

At the same time, country B does have a comparative advantage in computers (internal ratio for computers of .83 [B] versus 1.13 [A]). Thus, when compared to making computers, producing cars is internally less costly for country A than for country B; whereas, when compared to making cars, producing computers is internally less costly for country B than for country A. Each country can make money by doing what it does best, since its internal ratio of costs is smaller than that of the other country's ratio of costs for one of the two items. In this case country A benefits by trading cars and country B by trading computers. In brief, in international trade a country has a comparative advantage over its trading partners with regard to an item if the country produces the item for less cost relative to other goods it produces.

Now, assume capital as well as goods can cross borders. Since country A makes both cars and computers for less total cost than country B, it has an absolute advantage in both cars and computers. As a consequence, country B's capital moves to country A, where it is more efficiently used to make a profit producing both cars and computers. The workers in country A have jobs; those in country B don't, while the people with mobile capital in both countries make money.

In a global market economy where trade is based on absolute advantage, the benefits primarily accrue to those with capital, because workers will have employment only if they are willing to accept a total compensation package (pay, health care, retirement, workplace safety, and so on) below that of workers in other places. If they don't, capital will move to where it can be used to yield the best return. Clean air and water, species and ecosystem preservation, and abating climate change are likely to be cost effective in the long term, but capital generally moves to maximize profits in the short-term. Since neoclassical economic theory affirms that capital is most effectively used in the place that provides the lowest total cost of production in the near-term and since resource preservation is a cost whose benefit is difficult to price yet requires up-front money, the environment too is compromised when capital moves to maximize the return on its investment.

When *For the Common Good* was published, Daly thought he and John Cobb might have overlooked something and that their conclusions might be challenged by other economists. Almost fifteen years have passed, and no economist has put forth substantive challenges to their conclusion that comparative advantage has been replaced by absolute advantage.

With nations and international organizations such as the IMF and the WTO championing free movement of goods *and* capital, free trade has become what is termed "economic integration," where capital goes wherever it can be used most efficiently to make money. Economic integration contributes to efficient use of capital and other resources, less expensive goods, and diversity of goods, and it tends to promote rule of law, awareness and acceptance of other cultures and values, and perhaps world peace because of these benefits and the fact that war disrupts trade and commerce.

Economic integration also has substantial negative consequences for people. Daly contends that the trend already under way will continue to equalize wages, with high manufacturing wages in places such as the United States, Japan, and Europe declining toward those in the third world; whereas third world wages will rise little because of the availability of abundant labor. Daly laments, "We have here the abrogation of a basic social agreement between labor and capital over how to divide up the value that they jointly add to raw materials. That agreement has been reached not through economic theory, but through generations of national debate, elections, strikes, lockouts, court decisions, and violent conflicts—that agreement, on which national community and industrial peace depend, is being repudiated in the interests of global integration. That is a very poor trade, even if one calls it 'free' trade."

Globalized trade does more than push wages toward the lowest common denominator. Daly notes it erodes the fundamental grounding elements of community by allowing anything that raises costs to be eliminated through market competition: health care, infrastructure to ensure public safety, and social security and unemployment benefits, along with protection and conservation of the environment—all are threatened by global trade. Daly and Cobb write, "Free trade, as a way of erasing the effect of national boundaries, is simultaneously an invitation to the tragedy of the commons. Few people would advocate unrestricted migration [of labor] because [most] can intuitively see the tragic consequences. Free trade and free capital mobility have exactly the same consequences for wages and community standards, but are widely advocated in the false belief that comparative advantage still exists and guarantees mutual benefit."

Within national boundaries, the inequalities produced by absolute advantage are, to some extent, kept in check by rules, regulations, laws, and long-established community customs and ethics. But this legal and community framework does not yet exist internationally. In fact, with globalized trade and GDP growth driving the global economy, community constraints on absolute advantage are severely compromised.

Daly believes we can abate some negative aspects to global integration by creating a global network of integrated communities that are self-sufficient to the extent possible. One dollar spent on local food generates between $2 and $3 of other local business, whereas $1 spent for nonlocal food mostly leaves the community. Self-sufficiency in food should be a first priority for every nation, and within each nation the distance food travels from soil to table needs to be as short as possible. And, as advocated by Helena Norberg-Hodge, industrial agriculture needs to be and can be replaced in large part by local organic agriculture. In the long term, however, we will need to create a community-based natural systems agriculture as envisioned by Wes Jackson.

Trade is good. It provides competition, variety, and things not available locally, and the overall quality of life can be enhanced. Trade combined with mobile capital (or labor), however, wreaks havoc on communities and nations; accepting "free" trade means citizens will be employed under less than ideal conditions and environmental health will be compromised, while isolation will deny citizens the benefits of trade. Daly believes tariffs are the simplest and most effective means to replace the loss of comparative advantage and thus to strengthen communities by not allowing low-priced goods and services—delivered at low cost by impoverishing workers and the environment—to out compete local products and services. He does not advocate using tariffs to protect inefficient industries but does think tariffs are needed to protect efficient national policies. For example, tariffs should be employed to enable a country, through various internal policies, to cover the costs associated with environmental, health, and safety standards. Tariffs also preserve a community's legitimate responsibility to provide livable wages and social security for its members, as well as to decide if it wants to retain the capacity for some particularly important economic activity to maintain self-sufficiency such as making tool-and-die machinery or growing corn.

The overwhelming tendency toward economic integration in a globalized world, as documented in Helena Norberg-Hodges's story of Ladakh, makes Daly's recommendations for national economic sovereignty appear unlikely. Yet, the persistent and broad-based protests at meetings of the WTO, Global Agreement on Trade and Tariffs, North Atlantic Free Trade Association, and World Bank indicate many have realized that Daly is right: comparative advantage has become a mirage and global economic policies cannot be based on something that no longer exists.

FREE TRADE OF KNOWLEDGE

Daly has given sound reasons for regulating the trade of goods, but he does not think that regulating knowledge is beneficial. Here he affirms Thomas Jefferson's dictum: "Knowledge is the common property of [hu]mankind." The battle over genetically modified organisms (GMOs) illustrates this debate. The economic concern over GMOs arises from the question as to who owns genetic knowledge, and who has the right to manipulate genes. However, in addition there is great concern about the real possibility of an environmental or a public-health disaster caused by genetic engineering. With regard to making knowledge a commodity, in the case of the GMOs we ask: although a rather minor human manipulation adds value to an organism or a part of an organism, does the agent of manipulation necessarily own that modified organism or part of an organism? Intellectual property rights and monopoly ownership of knowledge for a substantial period of time are championed by corporations and some individuals because they contend that the rewards associated with rights and ownership are considered necessary to stimulate innovation and the search for knowledge. Simple observation challenges this assertion. I and most of my fellow scientists worldwide have been perfectly happy and willing to work 50-, 60-, and 70-hour weeks for a salary and the opportunity to do science without ownership or rights to the knowledge we obtain. On the contrary, innovation and new knowledge in science would slow to a snail's pace if knowledge were not freely shared. Imagine if Watson and Crick had patented the structure of DNA that they discovered but didn't invent, and then charged a royalty for the use of that knowledge in any process or activity. Absurd! Yes, but this is exactly what advocates of rights to and ownership of knowledge seek to implement.

An alternative economics of knowledge is given by Daly: "Once knowledge exists, its proper allocative price is the marginal opportunity cost of sharing it, which is close to zero, since nothing is lost by sharing it. Yes, of course you do lose the monopoly on the knowledge, but then economists have traditionally argued that monopoly is inefficient as well as unjust, because it creates an artificial scarcity of the monopolized item."

Freely sharing knowledge, however, can be problematic for a tribal, territorial animal like *Homo sapiens*. The receiving "tribe" might use the shared knowledge to kill or enslave members of the giving tribe. So, perhaps some knowledge should not be freely shared, because the consequences of sharing it are morally unacceptable. Nuclear weapons and biological warfare agents are examples. At the same time the ethical, environmentally acceptable, and efficient use of natural resources, capital, and labor can all be increased by sharing knowledge. Daly concludes that the IMF, WTO, and World Bank might better serve their stated goal of improving human welfare by advocating freely shared knowledge instead of interest-bearing loans, foreign investment, and economic integration, which promotes absolute advantage that, in fact, poorly serves humanity and the environment.

ETHICS IN ECONOMICS

Western culture, with assistance from neoclassical economics, has embraced the "invisible hand" of market economics, first noted in an obscure passage by Adam Smith in his classic text, *Wealth of Nations:*

> Every individual . . . neither intends to promote the public
> interest, nor knows how much he is promoting it. . . . He
> intends only his own gain, and he is in this, as in many other
> cases, led by an invisible hand to promote an end which was

no part of his intention. Nor is it always the worse for the
society that it was no part of it. By pursuing his own interest
he frequently promotes that of the society more effectually
than when he really intends to promote it.

Today the free market as advocated by politicians and their economic
advisors operates on the belief that the unrestricted pursuit of self-
interest will also serve the public good. What has long been ignored,
however, is Smith's assessment of what enables self-interest to effect the
greater good. In *The Theory of Moral Sentiments* he cautions that self-
interest cannot serve the public interest unless it is constrained by the
moral force of shared community values. This ethical foundation on
which classical economics is based has been lost as modern economists
have sought to make their discipline a value neutral science like
physics.

Daly rejects the notion that economics can be value free and states,
"[G]rowth has been and still is our central organizing principle. That
is precisely our problem. We need a new central organizing principle—
a fundamental ethic that will guide our actions in a way more in harmony with both basic religious insight and the scientifically verifiable
limits of the natural world." He advocates the following: "We should
strive for sufficient per capita wealth—efficiently maintained and allocated, and equitably distributed—for the maximum number of people
that can be sustained over time under these conditions."

On the surface his statement is couched in economic terms, but its
goals are expressions of certain values. "Sufficient . . . wealth," is a
value-laden, or subjective, term that means far more than some number of dollars: it includes health care, safety, security in old age, educational and employment opportunities, a pollution-free environment, the rule of law, healthy ecosystems, the arts, recreational
opportunities, and so on. All members of the community are to share
this wealth and perpetuate it, not just for the present community but
also for those to come, as "over time" implies. Environmental limits
are also acknowledged in the phrase "sustained over time." This state-

ment calls for an economic policy rooted in community values that is the result of an ethical consensus within a community. Here we face the daunting challenges of Daly's life's work: planting one economic foot in the natural sciences, with the other rooted in ethics and the humanities.

It was early evening on the second day by the time I cleared the bumper-to-bumper traffic of the Washington, D.C., beltway and settled into my seven-hour drive home. Herman Daly and I had been in constant conversation for two days, covering much of the material in five of his books, which I had read in preparation for our meeting. As I reflected upon my encounter with Daly and his nontraditional appraisal of our economic system, I remembered his initial response to my request to visit so I could write about his life. He warned me that his life was rather ordinary but that his ideas were important. I suppose that we all consider our lives ordinary, but they really aren't.

He was, however, right about his ideas. Daly has been in the vanguard of those economists who accept that major elements of economic theory, policy, and practice are bankrupt. Over three decades ago he stood almost alone when he challenged mainstream economists to recognize that the larger natural world must eventually constrain economic growth. As the founder of ecological economics, Daly mobilized a revolution in economic thought. His policy recommendations have provided civilization a course that might lead to a steady-state economy in which the practice of economics will be consilient with the principles of the natural sciences, and the well-being of all life—including humankind—will be a fundamental goal. Herman Daly has given us a good idea of where to begin, should we muster the foresight and political will to work toward creating wholesome communities in a finite world.

EIGHT

Accepting Uncertainty

Stephen Schneider and Global Climate Change

> What fraction of people in the world couldn't change their mind about something they deeply believe? A significant number; perhaps more than half. Of course, people are not irrational in every area, only some. What you have to do is to try to find components of their intellect where they are willing to be reasonable and negotiate with you. It really involves a value system that has to be taught. And that value system is: argument and evidence can change my mind, my belief; I can be rational.
>
> STEPHEN SCHNEIDER

The scientific evidence of a warming climate is replete. Earth is about 1.0°F hotter than it was in 1900. The decade of the 1990s was the hottest on record. In the Northern Hemisphere, species on the land and in the water are moving north, but whole ecosystems cannot move. The seasons are changing: spring is coming earlier and fall later. The reproductive success of myriad species is in jeopardy because reproduction is based on finely turned cycles and processes. For example, species migration, the budding of plants and hatching of insects, and the reproductive activity of organisms must be timed just right for insects to have leaves to eat, for birds to have insects to eat, and for flowers to have pollinators. Climate scientists now predict that over the next century the climate will warm by at least 1.8°F, perhaps by as much as 10.8°F. This degree and rate of warming will push untold plants and animals to extinction as ecosystems come undone—some estimates indicate that a quarter of all species could be slated for extinction by 2050.

In the face of such knowledge, what is the right action for humankind to take? Should we attempt to halt or reverse human-caused climate change, or just adapt to it as it happens? People respond to this question based on their perception of the consequences and the presumed likelihood of those consequences. Each of us makes this type of risk assessment routinely, as climate scientist Stephen Schneider illustrates with a student exercise. Taking a quarter out of his pocket, Schneider guarantees that when the coin is flipped heads and tails have equal probability of facing up. Then putting a one dollar bill on the table, he asks, "Who will put their one dollar on the table and risk the outcome of a coin flip?" Volunteers are many. After a few flips, he changes the amount on the table to $50,000 and asks who wants to put his or her $50,000 on the table and play. Since nobody can afford to lose $50,000, no one plays. He replaces the $50,000 with a one dollar bill and takes another coin from his pocket, saying that the probability of either heads or tails landing face up is not the same with this coin, but he doesn't state the probability. He then asks, "Who is willing to bet their one dollar on a flip of this coin?" The unknown odds dampen people's willingness to play. We all understand risk assessment in this very practical way: how likely are the consequences and can we afford those consequences? At the core, choices about whether to try to abate or merely adapt to climate change are that simple, but getting to the core is anything but simple.

As early as 1896 the Swedish chemist Svante Arrhenius in his studies of climate warned that increased concentration of carbon dioxide in the atmosphere could change Earth's climate. Noteworthy people say and write many things, but most of their pronouncements pass into the night, never to be recalled. A rare few predictions slumber, only to be awakened by the light of another era. And so it was with Arrhenius's warnings about atmospheric carbon dioxide concentration, which almost a century later became the subject of an intense political debate. In *Earth in the Balance,* Al Gore, then a senator in the United States Congress, accepted the assessment of many climate scientists, including Schneider, and wrote that increasing carbon dioxide could cause substantial climate change, which would be disastrous for human civilization. He concluded that climate change, along with a host of other negative environmental trends, were issues of such great concern that the rescue of the environment should be the organizing principle of the coming century. When he became Bill Clinton's 1992 vice presidential running mate, his outspoken environmental advocacy made him a target for political opponents.

Gore, along with climate scientists, was belittled in 1992 by columnist George Will, who took an anti-environmentalist stance in the *Washington Post*. Will made a special effort to devalue the science of climate change when he wrote: "Stephen Schneider of the National Center for Atmospheric Research [NCAR] in Colorado . . . is an 'environmentalist for all temperatures.' Today Schneider is hot about global warming; 16 years ago he was exercised about global cooling. There are a lot like him among today's panic-mongers." Will believed that the scientific community was being misrepresented by the United Nations Intergovernmental Panel on Climate Change (IPCC) and others who warned of the potential seriousness of global warming. By alerting the public to the changed positions of key climate scientists, Will attempted to highlight the uncertainty of the science and to make Gore's alarm appear foolish. Will, unfortunately, was ignorant not only about climate science but also about the process of science itself.

In fact, Schneider had coauthored an article that stated that human activities might result in another ice age. In 1971 Ichtiaque Rasool, an atmospheric scientist with the National Aeronautics and Space Administration (NASA), and Schneider published a paper in the prestigious scientific journal *Science* that concluded that the cooling effect of aerosols (dust, sulfate, other light-scattering particles) would counter and dominate the warming effect of carbon dioxide. They wrote, "If sustained over a period of several years, such a temperature decrease over the whole globe is believed to be sufficient to trigger an ice age." The paper was an immediate scientific "celebrity"—people either praised or attacked it. The conclusion was controversial, and, as with all important science, the work came under scrutiny.

When he cowrote this paper, Schneider had recently changed fields, and this was his first climate paper. A year earlier, when he was finishing his Ph.D. in plasma physics at Columbia University, he heard the conservation biologist Barry Commoner say on Earth Day that climate can be cooled by dust and warmed by carbon dioxide. He didn't believe Commoner, so he enrolled in the only course at Columbia on climate, which was taught by Rasool. Schneider recalls, "Rasool and I got on great. I didn't know anything about climate science, but I could solve differential equations. So I asked all kinds of questions. He was an intuitive thinker but couldn't solve the equations, rather he had an understanding of the system—climate and atmosphere. Sometime into the course he said, 'Why don't you come work with me on the carbon dioxide and aerosol problem. I'll take a chance on you, if you jump fields.'" Schneider agreed because he realized that the dream of virtually free, limitless electricity generated through the processes of plasma physics was not going to happen in his lifetime, if ever. In addition, Schneider was interested in researching a field that had a strong, direct social-benefit component, which plasma physics didn't have.

Schneider joined Rasool's lab as a postdoctoral fellow and started solving differential equations. He didn't know what they meant, but as was his style, he learned the science along the way. The solutions indicated that dust would win out over the heat-trapping gas carbon dioxide and cooling would dominate warming. Schneider had solved the differential equations, but, as he says, "I was the programmer and new then. Rasool didn't understand the calculations so when the attacks on our paper flowed in, he said, 'You made the calculations. You go out and defend them.' I protested that I was just solving equations he told me to solve. His response was to the point, 'Better start reading.' So I read the then twenty-five or so climate modeling papers in the whole world worth reading. Now there are 2,500."

Slowly, through the back-and-forth conversations and data discussions of scientific discourse, Schneider learned that he and Rasool had been wrong. The calculation was right, but they incorrectly assumed that dust was globally uniform and would thus have a uniform cooling effect. Instead, it was regionally heterogeneous and concentrated in industrial areas. Nobody had published this, nor had the distribution been calculated, but people in the field had data. Additionally, Rasool and Schneider had assumed that the stratosphere didn't affect global temperatures, but they learned that it did. With the stratosphere factored into the heat-transfer calculation, the carbon dioxide heating effect doubled. By 1974 new studies had added chlorofluorocarbons and methane to the list of significant heat-trapping gases. These adjustments reduced cooling and raised heating so that the overall result was reversed—the effects of heat-trapping gases dominated those of atmospheric dust particles. Schneider published a paper in the *Journal of Atmospheric Sciences* that showed what was wrong with the earlier paper. Argument and new information had changed his mind.

In 1976 Schneider wrote in his first book, *The Genesis Strategy* (coauthored with Lynne Mesirow), that "climatic theory is still too primitive to prove with much certainty whether the relatively small increases in CO_2 and aerosols up to 1975 were responsible for this climate change [of a slight global warming]. I do believe, however, that if concentrations of CO_2, and perhaps aerosols, continue to increase, demonstrable climatic changes could occur by the end of this century, if not sooner." As it turns out, he was just about on target. New data, better climate models, and a closer agreement between observed regional and seasonal patterns of climate and the more realistic climate models convinced IPCC scientists in 1995 to conclude guardedly that climate change had been detected and that some warming is attributable to human activities.

Schneider's assessment in the mid to late 1970s was unequivocal once he realized that the effects of heat-trapping gases would dominate those of aerosols. It was a good bet that the planet would warm, at least initially, because of two established facts. First, certain gases—the major ones are water vapor and carbon dioxide—absorb infrared radiation coming from the surface of the earth and radiate it back to the surface. This retention of heat initially coming to Earth in the form of sunlight makes the planet about 60°F warmer than it would be without these gases. The planet's average surface temperature is about 57°F, but without heat-trapping gases it would be about −3°F. Second, carbon dioxide concentration is increasing rapidly. In 1975 it was 12 percent higher than in 1900 and in 2004 is over 30 percent higher than in 1900. This accelerating rate means that it is very likely that carbon dioxide concentration will double between 1900 and 2100, and, with this doubling, surface temperatures are also likely to increase. Schneider puts it this way: "I was quite confident twenty-five years ago that you can't double CO_2 concentration in the atmosphere and thereby add four watts per square meter of retained heat and have no effect. The probability of no effect was astronomically low to me."

In response to Will's September 3, 1992 column, Schneider immediately wrote him a detailed letter to explain why the columnist's accusations were misplaced. A person on Will's staff called Schneider to apologize, but Will didn't retract his column. Will's beliefs and public position would not be validated if he were to recant his column attacking Al Gore's climate change policy and explain Schneider's change of position. Schneider contends that "if you can't change your mind based upon new evidence and argument, you are a dangerous ideologue. You have got to be able to change your mind as you see the reason of alternative viewpoints. Otherwise, you are in deep trouble if the other guys hold an opposite ideology and nobody can agree to talk. When you have absolute values that are so strongly held, two outcomes are likely: subjugation, either you win or you lose, or violence."

As a student at Columbia University in New York City, Schneider experienced firsthand the 1968 riot and its aftermath, both of which imprinted on him the difficulties of resolving conflicts between people who hold unwavering ideologies. The backdrop was the Vietnam War with the Students for Democratic Society (SDS) facing off against right-wing, pro-war students. Schneider recalls, "I didn't want any part of either side. I was on campus when I saw the police coming. Sociologically we had lower-middle-class Irish and Italian cops coming to Columbia to deal with a bunch of rich white kids whom the police already resented because the students were out there challenging the belief systems of their parents who fought in World War II. This was going to be a bloodbath. I went home and listened to it on the radio. It was a bloodbath. What did you expect? The university handled it all wrong. It cost the president his job, but the riot provided an opportunity for restructuring the university."

The Columbia administration proposed that a student committee be elected to work with faculty on restructuring the university so that it could more effectively deal with complex and potentially inflammatory situations. One student candidate from the school of engineering stated that the students' job was merely to be students, which was enough to motivate Schneider to run for the committee himself. In his campaign, Schneider stuffed engineering school mailboxes with his one-page position statement of classic liberal philosophy: moderation, social responsibility, equity, fairness, and individual and group rights. He won by a two-thirds majority.

Allen Temple, then president of Chase Manhattan Bank and a Columbia trustee very much involved with the restructuring, invited the student committee to his office in the Chase Building on Park Avenue for informal discussions. Temple's office was ostentatious, and the first meeting did not go well. As the students were leaving, Schneider said to Temple, "Do you want to understand why people don't respect you? Will you come to my old fraternity, if I invite you?" Temple agreed to come. Again, the discussions went badly. The bank president wore a three-piece suit and did not get along with any of the students present. Neither side listened to the other.

Temple was pretty upset because he felt that the students had not shown him respect, and therefore he was reluctant to deal with them. The fraternity members felt the same way about Temple. SDS contended that the process was completely illegitimate and that the only way to deal with illegitimate authority was with violence. In the beginning of the discussions, conflicting ideologies on campus were so strong that there was no room for substantive discourse. Schneider approached Temple again, suggesting that perhaps they could get on with the process in another discussion. And he asked Temple to wear something else at the next meeting. They met a month later, and Temple wore a polo shirt, casual pants, and tennis shoes. This time the riot was further in the past and everybody was calmer. Schneider recalls, "We actually had a very productive conversation that day—talked about an agenda for progress. Then we brought those informal agreements back to the student committee with the rest of the trustees and worked toward presenting the proposal to the faculty senate. Everything was fine. In the end it was stuffy Allen Temple who was the one who compromised. That is where I learned you can negotiate with people who seem to be the enemy, if they exhibit flexibility. You need to have an open mind as Allen Temple did."

As the student committee worked with the faculty on restructuring the university, Schneider learned another important thing about people. He found that some very articulate professors, whom he greatly admired for how they could express the ideology he believed deeply, were egotistical and had questionable characters. There were others whose ideologies he didn't share but whom he respected for their human qualities, decency, and willingness to negotiate. As Schneider says, "I learned you better separate someone's character from somebody's ideas. They aren't always the same. It would have been so much nicer to live with the belief that people's ideas and character are the same, but they aren't. That was a painful but very important thing to learn."

When Schneider went to work as a research associate with Rasool at NASA's Goddard Institute for Space Studies in 1971, the Institute was run by Bob Jastrow, an extremely articulate scientist but an iron-fisted manager. Almost a year after the 1971 *Science* paper predicting another ice age came out, the *New York Times* ran an op-ed piece that sarcastically said Earth can't warm and cool at the same time, so don't worry—neither will happen. Schneider responded with a letter to the editor stating that the science is yet unclear as to whether Earth is warming or cooling, but a large temperature change is a problem because all life forms and human civilization are adapted to present conditions. He emphasized that global temperature change is a very serious matter not to be laughed at, but studied. The *Times* published his letter, noting after his name that he was at the Goddard Institute. The day it was published Schneider was visiting Will Kellogg at the National Center for Atmospheric Research in Boulder, Colorado. When Jastrow read Schneider's *Times* letter, he was irate because he felt that he alone should represent the Goddard Institute. He phoned Kellogg and asked, "Is Schneider there?" and then said, "Well, keep him. I just fired him." This was Schneider's first lesson about being a scientist in the public arena—your boss might not like it.

Rasool smoothed things over and Jastrow rehired Schneider. In the meantime, however, the postdoctoral fellowship that Kellogg had pledged to arrange for Schneider the day he was fired by Jastrow materialized. Kellogg and Schneider had met about a year earlier in April 1971 when Kellogg gave a lecture titled "Man's Impact on Climate." In spring 1971, Schneider had just begun working with Rasool, and, at the lecture, Schneider approached Kellogg to tell him about the dust–carbon dioxide calculations he was doing. Kellogg asked him to attend and to help write up a report of a three-week conference in Stockholm, which was being organized for the purpose of defining the emerging field of human impacts on climate. If Schneider wasn't already convinced, that conference settled it—he would become a climate change scientist. Schneider had enjoyed working with Rasool, but he couldn't pass up the opportunity to work with Kellogg and his colleagues at NCAR. Within months of Schneider's arrival in Colorado, NCAR began a major reorganization. He joined with several others to write a proposal for a program to study climate, including human influence on it, which was accepted and then funded by the National Science Foundation. Nine months into his postdoctoral fellowship, Schneider was made deputy head of the climate group at NCAR.

In early 1973, about a year after moving to NCAR, Schneider participated in a session on climate change at the annual American Association for the Advancement of Science meeting. At the end of the session he reversed an old Mark Twain quip, saying, "Nowadays everybody is doing something about the weather, but nobody is talking about it." Two days later, Walter Sullivan, a *New York Times* science writer, used Schneider's quip as his headline for an article on climate change. Sullivan's article was accurate and responsible, but when it was picked up by other newspapers across the country, in a few cases it was altered beyond recognition. One version even changed the theme of weather modification to water modification. Schneider relates a colleague's response to this distortion: "I got back to NCAR, and, on the door of the map room, the place where everyone congregates, is this absurd clipping with 'Bullshit' stamped all over it. I learned that when you go public, not only does what you say often get distorted, but also other scientists are hostile because you are supposed to earn your reputation by working twenty years hunched over the bench."

The hostile response to the national attention that Sullivan's article gave Schneider might have been predicted. The ink on his Ph.D. was barely dry and he had but two climate papers to his name, yet he had been quoted in the *New York Times;* and to some fellow scientists such attention may have seemed unmerited. Furthermore, then as now, interdisciplinary inclinations and an interest in science policy are not generally welcomed by fellow scientists because such tendencies often lead to actions that aren't considered science or what "real" scientists do. Shortly after the clipping had appeared on the map room door, Schneider learned from two senior scientists that he had been nominated for a prize awarded to young scientists, because he had discovered the influence of cloud height on climate stability. But another senior colleague cautioned, "Steve, you've just got to stop this public stuff, because people think it is not real science. It is going to cost you. They have an image of what a scientist is, and you are not fitting it."

Schneider didn't get the prize—perhaps he had lost out on such recognition by taking his science public—but he continued to envision a scientist's role as far more than just generating data for other people to use or misuse. After all, the general public and most politicians are not scientists, nor do they have any real understanding of science, how it works, or what scientists' pronouncements mean. Of course, politicians have staffs and their staffers may have some scientific expertise,

but will the staffers get the science right and what kind of policy recommendations will they make? Active participation in the political
process by accomplished scientists increases the probability of greater
accuracy in the presentation of science-related issues to both legislators
and the public. Even in his graduate school days Schneider had been
inclined to venture beyond traditional disciplinary boundaries, and
while at NCAR, he asked himself, how far would he go beyond the
traditional role of a scientist? His boss there, Walter Roberts, who was
then also president of the American Association for the Advancement
of Science, knew that Schneider would be exceptional at science policy. But he also recognized Schneider's immense potential as a scientist. Roberts counseled him to stick to science but assured him that he
would provide him with contacts that would enable him to nurture his
desire for broader involvement in public policy.

Schneider took Roberts's advice: he continued to do science but
also made use of his boss's policy-related contacts. In the subsequent
three decades he published over 200 scientific papers. More than a
dozen of them were seminal in establishing an understanding of climate change and in articulating concepts in the field and creating analytical tools that have substantially advanced the field. He founded the
interdisciplinary journal *Climatic Change,* which provided not only a
model for interdisciplinary research, but also peer review criteria for
the evaluation of interdisciplinary research. In 1992 he was honored
with a MacArthur Fellowship, also known as the "genius award," for
his science and "for his ability to integrate and interpret the results of
global climate research through public lectures, seminars, classroom
teaching, environmental assessment committees, media appearances,
Congressional testimony, and research collaboration with colleagues."
This recognition acknowledged Schneider's unusual talent for making
climate science, and the choices before us, understandable.

Earth's climate is the historic result of an unknowable number of inter-
acting, forever-changing factors that include: the sun's orbit within our
galaxy; comets; solar output associated with sunspots; the shape of

Earth's elliptical orbit, its tilt, and the timing of its closest approach to the sun; movement of continents on their plates; mountain building; volcanism; geomagnetism; the composition of atmospheric gas, whose most important components are water vapor, carbon dioxide, and methane; ocean temperature; albedo, or reflection of light by clouds, dust, ice, snow, vegetation, and ground and water surface; Antarctic and Arctic ice; ocean circulation, including the North Atlantic thermohaline system; sea level and glaciation; dimethylsulphides produced by marine algae that influence the concentration of atmospheric sulfates and thereby effect cloudiness; differential influx and efflux of solar radiation from equator to poles; atmospheric circulation; and many others. Clearly, climate is a complex adaptive system that is very much like an ecosystem. Because of the number and diversity of factors that determine climate, no scientist can predict with a high level of confidence Earth's climate in 2050, or for any time beyond a few years.

From history and paleoclimate records, we do know, however, that the climate will not stay the same. Thus, we have two challenges before us: first, avoid human activities that might result in substantial, rapid climate change, because all existing life forms and human civilization are adapted to the present climate and large-scale extinctions might result from such change, and second, learn how to adapt to and survive the climate changes that do occur. The first situation has commanded the attention of natural scientists, because many are worried about our ability to do the second. The assessment of the scientific community in general, and of climate scientists in particular, is that we know human activities are forcing climate change, but the timing and consequences of that change are impossible to predict with certainty.

The assessment that the climate is changing and that we are forcing the change comes from several interrelated lines of evidence. Historic temperature records going back more than a hundred years show the planet's average temperature has increased about 1.0°F in the past hundred years, whereas some places have warmed much more—the north slope of Alaska and areas in western Canada have warmed 1.8°F to 3.6°F per decade since 1976. Most of Earth's warming has come in the past several decades. Many documented changes provide indirect evidence for a warming planet: increased severity and frequency of heat waves; spread of diseases out of the tropics; changes in the timing of the seasons—spring earlier and winter later; sea-level rise and coastal flooding; coral reef bleaching; movement north of aquatic and terrestrial species' ranges in the Northern Hemisphere; pronounced Arctic and Antarctic warming; and increasing extremes in precipitation and associated flooding, droughts, and fires. All of these changes are consistent with a warming climate, and it is unlikely that all are random fluctuations. Taken together the direct and indirect evidence has convinced many scientists, and others, that the earth is warming. But why is it warming?

It is well established that carbon dioxide, water, methane, and other gases in the atmosphere cause the retention of the heat radiated from Earth's surface and that these gases serve as a blanket over Earth. Historic records and various types of analysis show that carbon dioxide has increased about 30 percent and methane about 150 percent in the past century. The IPCC, the U.S. National Academy of Sciences, and other government institutions and panels have assessed several decades of climate science research and conclude that human activities, primarily the burning of fossil fuels and cutting down of forests, have made a major contribution to these heat-trapping gas increases. The observed warming trend is consistent with the observed increases in these heat-trapping gases. The critical question, however, is whether the heat-trapping gas increases caused by human activities are responsible for the temperature elevation.

Scientists try to answer this question by creating and manipulating climate models. In these models scientists use the known natural laws of physics expressed mathematically in the form of differential equations and code them into computer algorithms that enable a computer to re-create conditions that mimic some aspect of physical reality, like the current weather in England or the present average global temperature. A model that replicates the present climate is then used to assess the future under modified conditions. Over the past few decades, the power and speed of computers have expanded rapidly, allowing quicker construction and easier manipulation of climate models. Today earth system models are able to incorporate the effects of many components of the climate system, including atmosphere, land surface, oceans, and solar input. And models can show transient changes as system components interact over time, with the incremental increases in heat-trapping gases and other inputs, much as it actually happens.

Over the past two decades various groups of scientists around the world have run numerous equilibrium simulations in which a parameter like carbon dioxide concentration is changed, and the model is run until the temperature stabilizes. A doubling of the carbon dioxide concentration in all of these models shows a fairly consistent outcome: the climate warms at the surface by roughly 3°F to 9°F. These simulations demonstrate that the basic physics holds true: change the energy balance by adding carbon dioxide and Earth's surface warms. The first Earth system models increased over time heat-trapping gases and sulfate aerosols (which lead to cooling), showing patterns of temperature change that predicted fairly well the actual changes we have already seen in the past decades. Current Earth system models that link more elements of the climate system provide "fingerprints" that show a definitive connection between heat-trapping gas levels and observed temperature changes with ever-increasing fidelity. Models predict polar warming will be several times greater than the global average, which matches actual observations. Since the major heat-trapping gas associated with this warming is carbon dioxide, the overall process by which this increase in temperature occurs is simply called the carbon dioxide blanket effect.

The validity of climate models for predicting future climates resulting from the human-mediated carbon dioxide blanket effect is reinforced by the fact that models have accurately predicted levels and duration of global cooling following present-day volcanic eruptions, as well as changes to the seasons, resolving nicely the different patterns in the Southern and Northern Hemispheres. None of the predictions is absolutely accurate, because models only take into account a limited set of factors among the myriad that create climate. But the approximate accuracy in a wide range of situations between recorded climate changes and the model-generated climate changes has given climate scientists ever-increasing confidence in the capacity of climate models to predict future climates.

Taken together, the evidence—increase in surface temperature, changes in many physical and biological patterns authenticate warming, the compatibility between the basic physics of heat retention in the atmosphere and the known increases in heat-trapping gases (largely mediated by human activities), and the validation by climate models of cause and effect between increased heat-trapping gases and a warming planet—has led to a consensus among climate scientists and others: Earth's climate is warming and human activities are a major cause.

With the exception of a few skeptics, and there will always be outliers, the scientific consensus is that Earth has warmed about 1.0°F in the past century and will very likely warm several more degrees over the next hundred years. This consensus leads to two questions: first, is this projected warming of concern? and, second, if we should be concerned, what should our response be? Natural scientists who are familiar with the data believe it is dangerous to persist in forcing climate change. The 1996 IPCC Working Group I, on which Schneider served, concluded its executive summary with:

> Future unexpected, large and rapid climate system changes
> (as have occurred in the past) are, by their nature, difficult to
> predict. This implies that future climate changes may also
> involve "surprises." In particular these arise from the non-
> linear nature of the climate system. When rapidly forced,
> non-linear systems are especially subject to unexpected
> behaviour. Progress can be made by investigating non-linear
> processes and sub-components of the climatic system.
> Examples of such non-linear behaviour include rapid cir-
> culation changes in the North Atlantic and feedbacks associ-
> ated with terrestrial ecosystem changes.

When the last ice age came to an end and ice retreated to the poles and into the high mountains, Earth warmed about 9.0°F to 12.6°F over a 5,000-year period, or an increase of about 2°F per millennium. The 2001 IPPC report concluded that if we don't reduce human-associated carbon dioxide emissions by 60 to 80 percent, global temperatures will increase between 1.8°F and 10.8°F over the next century. Even the 1.8°F increase in a century is ten times faster than the sustained average rise seen at the end of the last ice age. By continuing activities that substantially increase carbon dioxide emissions, we are forcing climate change at an alarming rate, thereby courting disaster. We have, in Schneider's words, "imaginable conditions for surprise. A surprise being when a salient community didn't think it was going to happen or didn't think about it. But there could be surprises that nobody has thought about, written about, or even gleaned." Scientists know that rapidly forcing change on a nonlinear system like climate will induce behaviors in the system about which nobody has thought. And the faster we push climate change, the more likely it is that unexplored, unimagined outcomes will occur. Our civilization, as well as all ecological communities on Earth, is adapted to the present climate, with its record of relatively consistent, predictable weather. The greater the magnitude and the more rapid the change, the more difficult will be the adjustments for humans and the rest of life. Some organisms will do better than others, but the overall probability of extinction will increase for many species. The cascading effects of extinctions are impossible to predict, but the resulting biological turmoil is likely to reduce Earth's carrying capacity for many species in particular and for life in general.

The possibility of unforeseen human catastrophes, along with the certain devastating effect climate change will have on biodiversity, greatly concerns the majority of natural scientists; however, the dire consequences of rapid climate change seem to trouble only a few economists. The value judgments of economists and of ecologically oriented natural scientists about the seriousness of climate change differ for a host of reasons, but at the core these groups hold two very different worldviews. As my conversations with maverick economist Herman Daly indicate, for the neoclassical market economists who guide today's global economy and whose guiding principle is efficiency, current market models and forecasts cannot easily accommodate catastrophic widespread surprises, which may or may not occur. Another important principle in economics is substitution. If one kind of fish or tree is no longer available, then we can use a different type of organism or find another way to meet a need for this one. Alternatively, ecologists contend that many aspects of Earth's biologically and physically created life support—for example, our present climate!—have no substitutes, or at least none that might be available or affordable. They look upon possible outcomes of a warming climate, such as a major shift of the Gulf Stream or the devastation of major elements of biodiversity, as changes whose cascading consequences humankind might not wish to face or modern civilization might be unlikely to survive.

First-rate neoclassical market economists such as William Nordhaus have seriously addressed the economics of climate change in an attempt to guide public policy. Nordhaus's pioneering analyses in the early 1990s created the Dynamic Integrated Climate Economic model (DICE), the first integrated assessment of economics and climate change. DICE assumed a doubling of carbon dioxide and the IPPC median value of a 5.4°F increase in temperature by 2100. By looking at the effect on the U.S. economy of the exceptionally hot year of 1988, Nordhaus concluded that climate warming would primarily affect agriculture, which accounts for about 3 percent of the gross domestic product (GDP). The 1988 heat wave in the United States reduced crop yield by about one third, so simplistically he postulated that we can expect a 5.4°F warming to decrease GDP by about 1 percent. Even if the GDP were to decrease by two or three times this amount, it would only be reduced by several percentage points. With these assumptions and incremental adjustments to them, DICE and its successor models have indicated that the economic costs of reducing the use of fossil fuels to prevent more than a small fraction of climate change exceed the economic benefits, as measured by the GDP, of maintaining crop yields.

Schneider accepted Nordhaus's assumptions, and, soon after the DICE assessment was published in 1992, he plotted the profile of economic growth according to the neoclassical growth model with and without fossil fuel–related carbon dioxide emissions reduction. He discovered that the then-recommended 20 percent cut in emissions spread over 100 years would slow the growth of the economy by perhaps 10 years in a century. In other words, the approximate 450 percent growth in per capita income predicted to occur from 1990 to 2090 would take until 2100 if emissions were reduced by 20 percent. The total value of actions required for emissions reductions over the century expressed in 1990 U.S. dollars was estimated to be $5 trillion. Nordhaus countered that it was too expensive and inefficient to spend the equivalent of the total 1990 U.S. GDP merely to prevent a several percentage-point loss in GDP.

After ten years of back-and-forth discussion and of gathering new information, Schneider and Christian Azar revisited the cost analysis of stabilizing carbon dioxide emissions at different levels by 2100, again using the neoclassical economists' growth model and their high-end cost assumptions. The estimated expenses over the century at today's costs, given in 1990 U.S. dollars, for reducing carbon dioxide emissions to achieve atmospheric concentrations in 2100 of 350, 450, and 550 parts per million (ppm) were $18.0, $5.2, and $1.9 trillion, respectively (carbon dioxide concentration was about 350 ppm in 1990). With global economic output in 1990 at about $20 trillion, these costs certainly appear prohibitive. However, if we consider the 2100 world GDP, predicted by Schneider and Azar using economists' assumptions, the difference in cost between doing nothing to reduce carbon dioxide emissions and stabilizing carbon dioxide levels at 350 ppm is close to negligible. That is, the growth of the global GDP to perhaps $250 trillion by then would be slowed by 1 to 3 years, depending upon the cost of emissions reductions (estimated range: 3 to 6 percent of global GDP) and rate of economic growth (estimated range: 2

to 3 percent per year). Thus, waiting until 2103, instead of 2100, to be 5 to 10 times richer than we were in 1990 seems like a small price to pay to avoid forcing large climate changes that appear likely if we maintain or increase present levels of carbon dioxide emissions. (This analysis is conservative in that it does not factor in environmental benefits that will certainly accrue with reduction in carbon dioxide levels and could make the actual cost to society as a whole extremely low, perhaps negative.)

Although this economic analysis, which accepts that the economists' assumptions that unchecked climate change will have minimal negative effect on the economy are correct, is sufficient to make emissions reductions seem economically palatable, it does not consider the potentially huge costs of surprises. For example, the Gulf Stream brings warm water from the tropics up the eastern coast of North America and to Europe, and is a major part of the global conveyor belt that circulates water between hemispheres. Wind-driven evaporation of the warm Gulf Stream as it moves north is critical to global ocean circulation. Evaporation makes water denser, because it raises the salinity of the water, causing it to grow heavier and sink below the surface. As the Gulf Stream heads north, its increasing density eventually causes it to sink deep in the North Atlantic, where it then flows south as bottom water to eventually begin the cycle again. If sufficient fresh water is introduced into the Gulf Stream in the North Atlantic, the water will become diluted and will not sink, and ocean circulation will assume a different pattern. This shift could significantly and rather rapidly change climate, which is thought to have happened earlier in Earth's history.

Beginning some 12,000 years ago, after the end of the last ice age, the 500-year cold period that engulfed northeastern Canada and most of Europe appears to have resulted from a shift of the Gulf Stream. The cause of the shift seems to have been a sudden large influx of fresh water into the North Atlantic from melting ice in North America. During the previous interglacial period some 130,000 years ago, at which time Earth was about 4°F warmer than now, several flip-flops of

temperature of about 9°F occurred in Greenland, according to pre-
liminary research findings. These rapid shifts occurred in time spans of
decades and correlate to increases and decreases of carbon dioxide in
the atmosphere. As was the case 12,000 years ago, a changing pattern
of the Gulf Stream is thought to be the prime cause of the abrupt temperature fluctuations in Greenland.

Recently, oceanographers have detected broad, 10-feet-thick rivers
of fresh water flowing in the Labrador Sea of the North Atlantic. These
fresh water rivers are suspected to be melting Arctic ice, the result of a
warming planet. Fresh water tends to remain on top of saline ocean
water because of its lower density, thereby preventing evaporation of
the saline water below, but eventually it mixes with and dilutes the sea
water, making it less dense. These two consequences of fresh water
incursion may explain the freshening of the Labrador Sea over the past
several decades.

Will this large influx of fresh water from melting icecaps cause the
Gulf Stream to flip into another pattern that curtails its flow into the
North Atlantic? Nobody is certain, but this change appears likely
enough to consider the consequences—and soon. Climate history
indicates a flip can happen in as short a period as ten years. If this were
to happen, temperatures east of the Mississippi River in the United
States could drop by 4°F with the northeastern United States, northern Europe, and even northern Asia becoming even cooler. But the
fact that the rest of the planet is in a warming pattern makes accurate
predictions problematic.

Economists haven't even begun to calculate seriously the costs of an
altered Gulf Stream, but crude estimates of the annual price tag by
2100 indicate $2.5 to $62.5 trillion for lost agricultural productivity,
heating, transportation expenses, and other adaptations to the
changed climate. And these costs do not factor in the devastation such
a flip would have on biodiversity and the life-support functions it provides. Organisms and ecosystems now forced by warming to move
north would suddenly be driven south. Roads and cities, industrial
activities, farms, and other impediments would block movement to

places that still present an appropriate climate range. The biotic tur-moil that would ensue does not bode well for maintaining the current level and kind of human activities in these regions.

The real possibility of unforeseen effects appears to necessitate a much wider range of costs than those traditionally envisioned by econ-omists. However, since the future is unknowable, how we respond now to our predictions of future climates is based upon our subjective sense of possible outcomes and their costs, informed, hopefully, by scientific research of past climate events and predictions based on scientific knowledge. Nordhaus and neoclassical economists in general do not deny that climate change is likely, but they are perfectly happy to do lit-tle to avoid it, because in their view the economic costs are likely to be relatively low. They hold this view for two fundamental reasons: first, economists believe that products traded in markets would be minimally affected by warming and that nonmarket resources like species and ecosystems have little value; and, second, they believe that any particu-lar life-support feature has a substitute or can be replaced by human invention at low cost. Like economists, most natural scientists see climate change as likely, but their judgments lead them to advocate substantial expenditures to lessen the likelihood of major climate change. They believe that the damage is potentially huge to nonmarket resources that, in their view, have immense value and are essential for the viability of market activities. Schneider doesn't see the gap between these worldviews closing any time soon.

Schneider is, however, sanguine about resolving the differences between scientists and economists in the long term through education: "We teach interdisciplinarity well at Stanford. Everyone who graduates from our program, whether his or her specialization is biosphere, geosphere, or anthrosphere, has had all of the other classes. That per-son has to be literate in the physical, biological, and social sciences." In this environmental science program the students learn that the answers are not to be found within any one discipline and that open, respectful discourse can produce common ground for consensus. Schneider readily acknowledges his debt to economists who have convinced him of "two inviolate truths. One is opportunity costs and the other is trade-offs. Every time we choose to protect one thing, then we don't have those resources to do something else. We cannot duck opportunity costs. If we spend on a national park, then there is less for health care. We have to make very tough trade-offs. So I don't have a problem with economists who are trying to come up with some analytical rationalism for assessing opportunity costs to help us make appropriate trade-offs."

Economic analysis, however, can be misguided. Ascribing a pure economic value to biodiversity, for instance, is a poor basis for protecting species and ecosystems. It is important to know how much it costs to cut down a forest in terms of flood damage, water purification, and so on, but fundamentally species and ecosystems have enormous value beyond the purely economic. Schneider contends that "you protect species because there are 3.5 billion years of coevolution between climate and life that led to that species. How dare one species, hell-bent on getting more of us richer faster, wipe out 50 percent of everybody else when, if we slowed down, we wouldn't cause that kind of damage. It is arrogance. In my view, it is unethical behavior. But that's not a value shared by people who are stumbling over each other to have a fancier suit, a bigger car, two houses in the country, and holiday lavishly in faraway places. They have been mis-educated. They have been educated to keep their eye on the prize, the holy dollar grail, rather than the prize being a sense of continuity with the coevolution of climate and life."

Although the gulf separating neoclassical economists and climate scientists is a major stumbling block to implementing policies to avert or slow forced climate change and mitigate its effects, a similarly contentious rift that has also impeded substantive climate policy exists among natural scientists. Like economics in the twentieth century, over much of the same period the natural sciences held physics as the gold standard for science. In the physics model, quality science not only produces theory that describes with fidelity how the world works, but also enables accurate and precise prediction of future outcomes.

When a system is relatively simple and over time conditions remain essentially the same, like the movement of the moon and Earth, the physics model works well. We can describe the present and future relations between these two bodies accurately and with great precision.

Throughout most of the twentieth century scientists were inculcated with and adhered to predictability as an absolute criterion for quality science despite the impossibility of doing so in some areas. For example, our capacity to predict evolutionary outcomes outside of highly simplified laboratory experiments is nil, yet the body of evidence supporting the theory of evolution is as substantial as that for the theory of gravity. We speak of the law of gravity, but not of the law of evolution, I suspect, because of adherence to the criterion of predictability. Although many scientists believe quality science requires a process to be replicable so as to allow precise predictability, especially those in traditional disciplines in which this well-defined method works, an ever-increasing number of other scientists have accepted that such prediction is not an absolute criterion for quality science.

Evolution, ecology, and climate science are among the scientific areas of study poorly served by predictability as a requisite criterion for their validity. Because they represent complex adaptive systems composed of innumerable elements that are also a product of unique historic events, precise predictions of future outcomes are impossible. Edward O. Wilson contrasts the ease with which physicists can make limited predictions for simple systems with the difficulty of doing so for complex ones: "Physicists can chart the behavior of a single particle; they can predict with confidence the interaction of two particles; they begin to lose it at three and above. Keep in mind that [climate change] is a far more complex subject than physics."

If we cannot predict with confidence Earth's future climate, are all possible climates equally likely? Absolutely not! Some temperature changes are more likely than others. For example, climate scientists would give a very low probability to Earth's climate being 5°F cooler in 2100 and a high probability to its being 2°F warmer then. But these are subjective judgments based upon the scientists' assessment of current data and understandings. As climate scientists assemble more data, their assessments and their predictions will have greater validity and hopefully an ever-increasing influence on public policy.

Uncertainty about future climate can be reduced, but not eliminated. However, scientists' assessments of new data, along with their knowledge-based predictions, will enable government and ordinary citizens to make more informed policy choices. Already current analyses allow us to know that forcing rapid change on nonlinear systems is best avoided. When scientists recommend specific policies, however, their recommendations are influenced by personal value judgments, as well as scientific conclusions. Thus, when communicating with the public, scientists need to make a distinction between presenting the results of scientific investigations and evaluating possible risks of human actions on the one hand and advocating policies that they believe address the risks and thereby will make the future better on the other.

At a time when so many policy decisions require scientific understanding, many scientists are uniquely qualified to be at the forefront of the public debate or to champion colleagues who have entered the fray. But, by and large, most shun such activities and exposure. Science's cultural tradition admonishes scientists who do venture into the public arena, for a scientist's job is to establish what might be "true" by demonstrating what is false. If attempt after attempt to falsify a hypothesis fails, over time it becomes accepted as true, but conditionally so. The actual process of science is contentious and filled with nuance and detail. It is repetitive, often boring, and replete with more blunders and worthless outcomes than most scientists wish to admit. In a word, it is messy. The outcomes that get neatly packaged and eventually appear in peer-reviewed literature and then in textbooks, if they stand the test of time, reflect little of the process that created them. They do, however, still carry with them myriad assumptions and conditional parameters that must be accepted as true for the conclusions to hold. The public has no patience for such detail, and the scientific community provides no rewards for communicating science to the public—discovery is the only grail. In fact, as Schneider and many others who have "gone public" with their findings have learned, peers can be hostile and jobs and prizes can be lost.

The public venues for popularizing science and advocating a particular understanding of it are severely constrained by the limited time and space given them by mainstream media, as well as by the brief attention span of the general public. These limitations pose a dilemma for the scientist who ventures into the public domain: it is a free-fire zone, because once a scientist goes there, he or she becomes vulnerable to the criticisms and retributions of both colleagues and the public. Schneider has characterized this predicament as the double ethical bind: as a scientist, he or she must honor the scientific method, in which all uncertainties and assumptions are to be communicated. Yet as a citizen, the scientist seeks to better the world by championing a particular worldview. For example, Schneider advocates that we curtail drastically the burning of fossil fuels to avert forcing climate change; botanist Peter Raven and ecologist Paul Ehrlich work tirelessly to promote preservation of biodiversity so as to maintain Earth's life support. To get media coverage, the message needs to be simple and attention getting, so the nuance and details of the science must be eliminated. Schneider sums up the challenge this way: "The double ethical bind for communicating science to the public, then, is for the scientist to find appropriate balance between being an effective agent for change and being honest about the limitations of the state of knowledge."

Achieving this balance is an awesome feat that takes talent, skill, and effort, even for masters such as Schneider, Ehrlich, and Raven. So, it is little wonder that the vast majority of scientists remain in their laboratories, stepping out only to question the veracity of their few colleagues who do venture into the public domain. Unfortunately, the result is a scientifically ignorant public, because at the extremes scientific information remains within the science community or is poorly presented to the general public. Schneider sees raising public awareness of scientific issues as the only responsible thing to do: "I believe it is, quite frankly, elitist to suggest that we in science are too lofty or too busy to take time to explain things honestly and simply to the nonspecialist public which pays for most of our activities." Some professional organizations, including the Union of Concerned Scientists and the Ecological Society of America, agree and in the past decade have initiated programs to train and support scientists as advocates of a scientifically grounded worldview of the contentious issues of climate change, biodiversity loss, population stabilization, economic growth, pollution, overconsumption, energy sources and use, and others.

Stephen Schneider's career also illustrates a changing attitude among the scientific elite. For example, the astrophysicist and popularizer of space Carl Sagan was never elected to the U.S. National Academy of Sciences (NAS) even though he was nominated several times, apparently because his public advocacy of scientific literacy with regard to space, through television and other popular media, violated the code of conduct for scientists. Sagan's death in 1996 focused attention on this prejudice within the scientific community, because many felt his scientific contributions certainly would have merited his election to the NAS. In contrast, despite three decades of outspoken and often contentious public advocacy on issues of global change, Schneider was elected to the NAS in 2002. This highlights an important shift in the acceptance of scientist-advocates by science's elite.

Schneider's election to the NAS despite his outspoken advocacy may encourage some scientists to go public, but such changes are glacially slow in the laboratories, professional meetings, and science classrooms where most scientists ply their trade. My own experience is typical. In order to draw attention to the problem of human overpopulation and its erosion of Earth's life-support capacity and numerous other ramifications, at the institution where I teach I organized a full-day event to recognize and educate the community about the meaning of the arrival of the six-billionth person on Earth in October 1999. When I sought support for the event from my department, I was publicly chastened by my chairman and several other faculty for advocating population issues. The department did not support the event, and only one of a dozen biology colleagues stepped forward to help. At the event itself, just four science faculty out of more than a hundred in the school of science made an appearance or gave any support, and they were all personal friends.

In both science and in the larger culture an increasing willingness to acknowledge and address environmental issues is detectable, but the inertia of established patterns of behavior will not be easily overcome. This assessment raises a critical question: what is our cultural capacity for change? Clearly a particular culture is not inherent to human beings; however, we cannot wipe the cultural slate clean and start anew. If our culture, which is eating itself out of house and home, is to survive, a metamorphosis will have to occur. Although humans can probably adapt to any one of the myriad problems we face, as Schneider writes, "[B]y far the most serious environmental [problem] of the twenty-first century will not simply be habitat loss or ozone depletion or chemical pollution or exotic species invasions or climatic change by [itself], but rather the synergism of these factors."

What are we going to do? When I catalogue humankind's most serious problems—from pollution and loss of biodiversity to industrial agriculture and globalization to a lack of social justice and overpopulation to climate change—I arrive at one conclusion: the time has come for our beliefs to change radically and, as a result, for our approach to living to change completely. We need policies based on empirical evidence and sound scientific argument, thereby enabling our social and economic systems to be consistent with the fundamental principles of nature. So it comes down to this: is it possible to metamorphose an economically centered culture into an ecologically centered one on a worldwide scale and in a way that accommodates human nature and behavior? Can we transform our current relations with each other and the rest of life so that we can reverse the course of diminishing Earth's life-support capacity that we have imposed upon ourselves? I don't know; however, the answer might be "yes" if enough of us emerge from our cocoons and use our new wings to take flight—keeping in mind the story of Icarus.

NINE

Ecological Design

David Orr and Education

> We need a perspective that joins the hard-won victories of civiliza-
> tion, such as human rights and democracy, with a larger view of
> our place in the cosmos. . . . When we get it right, that larger,
> ecologically informed enlightenment will upset uncomfortable
> philosophies that underlie the modern world in the same way that
> the Enlightenment of the eighteenth century upset medieval hier-
> archies of church and monarchy.
>
> The foundation for ecological enlightenment is the 3.8 billion
> years of evolution.
>
> DAVID ORR, *The Nature of Design*

Caterpillars are hardwired for growth: Evolution designed them to be voracious eating machines with little else in their behavioral repertoire. If eating were all caterpillars ever did, they would eat themselves out of house and home. But evolution has programmed them to transit from this growth phase. As winged adult butterflies they are past their growth phase and eat only enough to maintain themselves. In addition, along with a butterfly's zero rate of growth, the throughput of energy and material needed to sustain it ceases to increase. In Herman Daly's economic language, the adult butterfly "economy" has reached a steady state.

In contrast, humans have created a global economy with an insatiable growth agenda and no mechanism to shift to a steady state. Our modern agricultural system puts 1 calorie of food on our table at an expenditure of at least 13 calories, all the while degrading soil and polluting the environment. Our industrial society generates toxic waste that infiltrates our food and water supplies and creates health hazards. Our dependence on fossil fuels promotes climate change. Our numbers and degree of consumption push much of the rest of life toward extinction, thereby impoverishing our support systems and future choices. However, since the negative consequences of our choices seem distant, we continue with the perhaps foolish hope that the future holds solutions to all our complex problems. As a society, we are led to believe that we can go forward without making any changes, and everything will take care of itself—somehow.

Since the mid-1980s, in the United States an overwhelming majority of students entering college say that "being very well off financially" and "being able to get a better job" are primary reasons for seeking a college degree. The underlying assumption is that a better job and financial wealth will enable them to obtain more material things and the "good" life. To help students fulfill these aspirations, colleges and universities have honed their educational focus to serve this role of producing workers whose primary goal is to be consumers. Curricula that attempt to holistically educate students, but give them no particular advantage in the job market, command little attention and few resources.

The situation looks dire—the adjustments required to make the transition into the next century with global civilization intact demand formidable educational reform. Will higher education acquire sufficient vision and courage to take on the daunting environmental challenges of the twenty-first century?

Intermittent rain and a biting-cold wind chilled me as I walked to take my seat at the dedication of the Adam Joseph Lewis Center for Environmental Studies in September 2000. The rain then poured down as David Orr, chair of Oberlin College's Environmental Studies Program, lavished his praise and thanks upon dozens of people who had contributed to the creation of the Lewis Center. The building that houses Oberlin's Environmental Studies Program may be, for the moment, the "greenest" classroom-office building on any college or university campus, and certainly it is among the most environmentally appropriate buildings in the industrialized world.

During the two-day inaugural ceremonies, I listened to many speakers praise the Lewis Center. What they said made the impossible seem possible: a building that was constructed with either recycled or sustainably produced nontoxic materials, that purified its bathroom effluent and other liquid wastes on-site to be used again to flush toilets or water outside vegetation, and that would produce all of its energy while not releasing heat-trapping gases into the atmosphere. As the event progressed, sunlight streamed into the atrium of the Lewis Center. In that idyllic, sunny environment, with so many people celebrating the design ingenuity and environmental solutions of the center, I was lulled into believing that our most egregious environmental problems are at least manageable. Then Robert Krulwich, the moderator, asked, "Can we have six billion people with these new technologies?" The idyllic scenario withered into an image of dry, parched scree.

Several months prior to the Lewis Center dedication I had chatted with David Orr in Oberlin's Black River Cafe, a restaurant that was established a few years earlier by a graduate of the Environmental Studies Program and that serves local, organically grown food. Orr, pausing between bites of his sandwich, commented: "Nothing is moving fast enough and there is a lot of complacency in the United States. I do not have access to public policy. I am not in Congress. Periodically I do things in Washington, but my focus is my own backyard. That's it! And so the part of this ecological Berlin Wall I want to tear down is here on campus. We have half a dozen big projects going on around us. I don't know what it all amounts to. It is a drop in the ocean, as far as I can tell. But it becomes like that little butterfly, you know, chaos theory. Pretty soon you have a tornado of change coming because of what is happening in colleges and universities. Does that amount to enough? I doubt it. But that's what's in my power to do."

Orr is a modest man. Yes, he has plenty of ambition and vision, but at the same time he is modest. He is less concerned about getting credit for a project than about seeing it get done. His conversations and writings are filled with credit to others. He rarely, if ever, talks about himself or what he has done, and, when he does so, he uses the plural pronoun "we." It's the agenda, the dreams, and the agents of change that fill his conversation and writing.

I have watched Oberlin's environmental studies major become one of the most populated on campus. Its beginning can be traced to the late sixties and early seventies as anti–Vietnam War student protests diversified to include concerns for the environment. David Egloff, a biology professor, supported a group of students who established an environmentally focused dorm. Although the dorm initiative could not be sustained, student interest in the environment remained keen. It took until 1978 before a group of faculty created an interdisciplinary course on the environment, and, in 1980, Egloff became director of the nascent program. Environmental studies, however, remained a minor item on Oberlin's academic menu. Then, like now, environmental issues were not a priority in higher education's agenda, because the Academy mirrors society.

By the eighties, as Reagan-era policies eviscerated the Environmental Protection Agency (EPA), as well as many of the conservation and renewable energy programs spawned by the oil crises of the seventies, most environmental studies programs that had taken hold in the sixties and seventies were falling on hard times. In contrast, during the Reagan years Oberlin's environmental initiative survived and was nurtured by Egloff and his successor, Harlen Wilson. But neither director nor Oberlin had groomed an in-house successor with the passion or vision required to build the program. Where could Oberlin find a person who had the credentials demanded by an elite liberal arts college and who also had the foresight and energy required to establish legitimacy and prominence for its environmental program?

David Orr grew up among the rolling hills of western Pennsylvania where his dad was a minister and president of Westminster College. Some of the young Orr's spare time was spent at a cabin in pristine forest close to the Allegheny River. In relating this history to me he said, "I don't think I ever had a 'road to Damascus' sort of conversion to environmentalism. I think what happened in my case was I grew-up with a family that was not environmentalist, but my dad was an outdoorsman of sorts and my mother definitely was an outdoors enthusiast. I think I was imprinted early on by just playing in the woods and being around an impressive old-growth hemlock grove with incredible rock formations. And the result is what Rachel Carson describes in her little book *The Sense of Wonder*. You have to have some pegs in the mind to hang concepts on, and those pegs are formed by experience. When I began to read people like Rachel Carson and René Dubos, I had an 'aha' feeling. I knew what they were writing about and why."

His sister and brother were exceptional students, salutatorian and valedictorian respectively of their high school classes, but Orr stated, "I was never a student like that. I was always playing baseball or bas-ketball or reading." And, like so many synthetic thinkers, Orr is an avid reader. This started very early but solidified when he contracted a rare lung disease and spent some of his seventh- and eighth-grade school years bedfast. Without much to do, spending day-in and day-out in bed, he developed what would become a life-long habit of vora-cious and passionate reading.

About half-way through college Orr got seriously interested in learning. He went to graduate school for a Ph.D. in political science at the University of Pennsylvania. The excitement for Orr, however, was not in the political science department but in landscape architecture, where Ian McHarg was formulating ideas for his classic interdisci-pli-nary book, *Design with Nature,* and in anthropology where Loren Eiseley was integrating the human experience into an evolutionary matrix. For Orr, the ideas of these intellectual giants resonated with his early experiences in nature, laying the foundation for a deep environ-mentalism. He finished his Ph.D. in political science, but he said, "It was pretty clear that my passion had to do with living systems and the way human institutions intersect with these systems."

In 1971, as a new faculty member at Agnes Scott College in Decatur,

Georgia, Orr began to forge his environmental perspective. As a board member of Georgia's Nature Conservancy, he met Eugene Odum, a systems ecologist. On Conservancy walks in the woods and along the Georgia coast, Orr learned his ecology from this master. Around the same time, Orr began to read Herman Daly, the founder of ecological economics. He got a copy of a talk Daly gave at Yale, and his response was, "Hey, bingo! This guy is doing the kind of economics I think is important." *The Limits to Growth*, a landmark book by Donella Meadows and colleagues about impending constraints on human activities, came out in 1972. The next year Orr and a colleague at Agnes Scott raised fifteen thousand dollars to sponsor a two-and-a-half-day symposium that would bring in Herman Daly, Ian McHarg, Ralph Nader, and a dozen other forward thinkers. The event was a big deal, and newspapers and television and radio stations covered the event.

He earned tenure at Agnes Scott, but the political science department at the University of North Carolina at Chapel Hill offered him a nontenure-track faculty position with the promise of a tenure-track slot. David—together with his wife, Elaine, and their two boys—packed up and moved a bit north. At the same time, other things were brewing. Orr and his brother, Wilson, who was in the construction business, had been inspired a few years earlier by E. F. Schumacher's *Small Is Beautiful* to imagine their own experiment in environmental education. His brother took the lead in seeking a possible site for their environmental education center, while Orr presented the political science elites at Chapel Hill with his contrarian political science perspective.

There, Orr's courses on environmental politics were instantly oversubscribed, and the students were turned on by this unconventional political science teacher. Most of Orr's Chapel Hill political science colleagues were, however, cognitive aliens to his environmental worldview and couldn't fathom his perspective. And, on arrival, Orr was naive about how those in power respond to the unknown. The memory of those days still evokes passionate discourse from Orr: "I came in talking to my political science colleagues about issues that are long term, value-laden, and incredibly complex. These issues require learning first how the world works as a biophysical system and then how our discipline might apply to that reality. My colleagues found what I was doing so inconvenient and incomprehensible, even though it resonated with students and in peer-reviewed literature in political science.

"In the context of departmental business, I raised these issues only once in a memo to the political science faculty asking whether these issues are important. If so, in what way are they political? And if they are important and in some way political, how do we accommodate them within political science? The silence was deafening."

Reflecting further, Orr commented: "Those in charge influence outcomes by avoiding things they don't want to talk about. They don't respond; they just ignore. In a place dedicated to reason, reasons ought to have been given. The problem is disguised as academic rigor, which allows the powerful to make their critics invisible."

Speaking about his colleagues, Orr went on to say: "They didn't know enough, or care enough, to cross the chasm into an ecological worldview. The problem is simple dishonesty. The discipline is gerrymandered around uncomfortable things—value questions and ecological issues. You have got to start with how the planet works and what it can handle from us. We need the equivalent of a Copernican revolution that removes humans from the center, or a Darwinian perspective that puts us somewhere in the larger scheme of things. Such views still operate at the periphery of education."

In my view, ecological thinking remains at the periphery because our institutions and rituals, religious and otherwise, tell us wrongly that we are exempt from the limitations that apply to the rest of life. We may teach Darwin's truth about the evolution of life on Earth and recite with fidelity the observations and conclusions drawn from those observations, yet few of us live as if we believe that evolution has an effect on contemporary humanity. We just do not conduct our lives with evolutionary and ecological principles at the core. Despite being exposed to these principles, we grasp them, if at all, at the most rudimentary level. Colleges and universities—in fact, the whole educational enterprise—

share blame for this ignorance because many educators are deluded about the objectivity of their knowledge. Too often they profess knowledge without examining the evidence—they perpetuate received knowledge and purported wisdom without seriously questioning their validity. Is this judgment too harsh? Perhaps, but despite our culture's beliefs and actions to the contrary, overwhelming evidence favors the fundamental correctness of evolutionary and ecological principles. Orr tells a most relevant story about educators.

In one of his North Carolina courses, Orr had the students read E. F. Schumacher and Herman Daly. The scholarship of these economists is based on holistic analyses that establish the pervasive flaws in contemporary economic theory and practice and recommend in their place a human-scale and natural science–based economics. The faculty in the economics department at Chapel Hill openly dismissed this scholarship but did not explicitly criticize it. Orr recalled, "Standing on the sidelines of the soccer field while the kids were playing soccer, I turned to the chairman of the economics department, whose kid was out there, and I said, 'Jim, I know you don't like the work of Herman Daly and E. F. Schumacher, so how about coming into my class and giving a critique?'

" 'No, I'm too busy, can't do it.'

" 'Well, could we change the time?'

" 'No, I'm too busy.'

" 'Could we come to your office?'

" 'Too busy.'

" 'Could you just record some things for us and I'll play it in class?'

" 'No. No time for it.'

"So I said, 'Well, Jim,'—I took out a piece of paper and pencil— 'could you just write out something you think is wrong with Daly and Schumacher?'

" 'Oh, God damn it, David, I haven't even read those people!' "

Meanwhile, Orr's brother had been exploring possibilities for establishing their independent, experimental program in environmental education. In the late seventies, he found a suitable place they could afford—fifteen hundred acres in northwest Arkansas. David's time at Agnes Scott and North Carolina had convinced him that the conventional classroom was inadequate for environmental education. More was needed, but just what was necessary had to be determined.

David and his wife, Elaine, sold their home and put every dollar they had into Meadowcreek, the not-for-profit, 501(c)3 organization the two Orr families had founded. His brother's family did likewise. David Orr told me that, at that moment, "If you had asked me for a ten-dollar loan, I couldn't have given it to you." Once again he was pursuing a vision, but this one would be life transforming. He explained, "The contrast for me of being at North Carolina and then moving to northwest Arkansas was so powerful that it imprinted more deeply in me than any other single thing in my life. The transition from being in a place where self-congratulation ruled to living in the fifth poorest county in the second poorest state in the union was for me an indelible and incredible experience. As I looked back from the periphery to the center of the society, things became clear. Just like people in the Peace Corps, I found that it was a period of dis-adjustment. It is such a disconnecting experience as you unplug from the materialism and the wealthiest society ever on the face of the earth."

When Orr went to Arkansas to establish Meadowcreek, he had not thought hard about educational pedagogy or how an institution might deliver culture-transforming environmental education. To complete a Ph.D. degree program and to be a successful faculty member demand knowledge of subject matter and methodology, not competence in teaching or a background in education. He had firsthand experience teaching at both a small college and a large university. With regard to environmental issues, neither institution nor their programs and faculties had impressed him; everyone had been indifferent. Orr recalled, "As an educator in the 1970s concerned about the environment, I was impressed by how little we in the academy connect disciplines with head, hands, and heart. The academy is populated by indoor, talking people. We assume that somehow, out yonder, it will all work out, but we don't have to make connections. Anything like holism, systems, or patterns is pretty alien as far as the environment is concerned. Anything that requires physical competence is undervalued because head and hands are separate."

The Orrs founded Meadowcreek as an educational experiment in how to live in that particular place. Five areas of activity emerged that would later become both educational programs and the subjects of conferences as well as research: solar technology, sustainable forestry, sustainable agriculture, rural economic development, and land management using applied ecology.

The Orrs lived on a shoestring as they created a place-based pattern of living. They raised cattle, started a sawmill, and produced their own energy and food to the degree possible. By 1983 Meadowcreek had accumulated enough money to start building an educational center. The Orrs built their own facilities—conference center, library, kitchen, and accomodations that could house up to about seventy-five people. The rule was: if it happened in Meadowcreek's 1500 acres, it was potential curriculum. Curriculum was not knowledge imported from outside, but something that the Orrs did to sustain themselves in that particular place. If an event could be held at a Holiday Inn, they didn't want to host it. Anything that broadly applied to the issues of sustainability became curriculum.

Semester internships that usually fit into one of Meadowcreek's five ongoing activities attracted students from more than a hundred colleges as well as returning members of the Peace Corps. Other programs included elder hostels, summer conferences for gifted and talented high school students, and January terms for college students. Foundation leaders came for a conference on issues of sustainability, whereas bankers came for a conference on climate change. Meadowcreek hosted and helped fund two conferences that led to the book *For the Common Good,* by Herman Daly and John Cobb. The background for these conferences and internships was a working community: the Orrs and about two dozen staff were responsible for a little construction company with a backhoe and bulldozer, a machine shop, a welding shop, a farm with sixty or so head of cattle, and a fire department, and for maintaining and running it all. Education and the life of this basically self-sufficient place were never separate.

What emerged was a vibrant, environmental-education program. Music from local musicians, mostly bluegrass bands, often accompanied events. Nature hikes, swims in Meadow Creek, farmwork, cutting firewood, and building things filled the lives of conference attendees and visitors. Many noted environmentalists, scientists, and politicians participated: Wendell Berry, David Brower, Bill Clinton, Hillary Clinton, Wes Jackson, Gene Likens, Amory Lovins, Donella Meadows, Stephen Schneider, and George Woodwell.

During Orr's first thirty-six years he had acquired a formal education not all that different from many of his peers, but in Arkansas he found himself on "a road less traveled." Orr summed it up this way, "The experience of going to the periphery of the society was absolutely essential. It changed everything. The eleven years in that valley in Arkansas were to me the most informative of my life." His conclusion to *Earth in Mind* focuses the meaning of this experience:

> Were we to confront our creaturehood squarely, how would we propose to educate? The answer, I think, is implied in the root of the word *education, educe,* which means "to draw out." What needs to be drawn out is our affinity for life. . . . Education that builds on our affinity for life would lead to a kind of awakening of possibilities and potentials that lie largely dormant and unused in the industrial-utilitarian mind. . . . How will this awakening occur? Scott Momaday (1993) put it this way: "Once in his life a man . . . ought to give himself up to a particular landscape in his experience, to look at it from as many angles as he can, to wonder about it, to dwell upon it. He ought to imagine that he touches it with his hands at every season and listen to the sounds that are made upon it. He ought to imagine the creatures there and all the faintest motions of the wind. He ought to recollect the glare of the moon and all the colors of the dawn and the dusk."

Meadowcreek was that experience for Orr. "Meadowcreek is the only place I dream about—it merged landscape and mindscape," he said.

Orr never abandoned his mission to transform higher education. The innumerable people—students, farmers, educators, politicians, businesspeople, and ordinary folk—and their institutions that became part of the Meadowcreek mission provided the grist for his educational inquiry. Orr had come to Meadowcreek with the troubling knowledge that in the standard classroom one could not conduct environmental education. He came to understand that when one combined head and hands in a natural, process-based setting, one could unite different elements, the feeling and the analytical parts of the brain. As he told me, "When you really get it right, you join disciplines with different aspects of the full person." As the years passed he came to realize it was not so much the problems *in* education that needed attention, but rather, the problem *of* education. In the same way that Wes Jackson had recognized that industrialized agriculture is a failure, David Orr came to see that the educational system has failed us also.

In 1987, at Hendrix College in Conway, Arkansas, Orr facilitated the first campus food audit. The audit investigated where the college purchased its food and the fate of the waste generated, as well as the consequences of purchasing from certain sources and of the institution's waste disposal methods. In addition, the audit recommended changes—purchasing locally grown food and feeding waste to local farm animals or composting it—that might make feeding students less destructive both environmentally and socially and might provide them with better nutrition. The idea spread like wildfire. Since then, hundreds of campus audits of food, energy, materials, water, and waste have been done across the country. These resource audits provide students, faculty, and administrators hard data on resource use and waste generation, along with evaluations of an institution's environmental impact. The audits opened the door for environmentally meaningful change. At Meadowcreek, hand and mind had created an environmental learning model needed everywhere. But Meadowcreek was isolated in a back eddy far from the mainstream of education—hardly the place from which higher education's environmental perspectives could be transformed.

Meadowcreek was a grueling operation for everyone involved, but most of all for the Orr brothers. Raising hundreds of thousands of dollars each year required many pre-dawn, four-hour trips over rough, winding roads to the Little Rock airport. Sometimes, before leading an all-day program, they had worked much of the night with the volunteer fire department. And the community's fences always needed mending. After ten years Wilson Orr left.

Through student internships and the college's January term, Orr was in contact with Oberlin. It was 1989 and Oberlin was looking for a professor of environmental studies. Orr appreciated Oberlin as an exceptional institution with a tradition of challenging conventional wisdom. Here was a unique opportunity to launch his ideas into the educational mainstream at an institution small enough to change, yet large enough to be credible.

Orr considers leaving Meadowcreek the hardest decision he and Elaine ever had to make. He recalled, "Things were going exceptionally well and we had absolutely fallen in love with the place. I really felt close to the natural world there. We had forged out of that land a farm, a wood business, and an educational center. Although it was remote and physically demanding, I wouldn't take anything for that experience. Everybody ought to step out of their profession and then look back. Leaving it was a very, very tough decision. I really miss the connection that I had with nature at Meadowcreek."

David Orr returned to academia with high hopes, but he found that little had changed. He understood that the lack of connection to the natural world estranged campuses and academic communities from their ecological roots. How could Meadowcreek's holistic educational model that addressed the problem *of* education be reestablished in education's mainstream at Oberlin? Would Oberlin, true to its motto, "Learning and Labor," be open to educational venues beyond its buildings' isolating walls?

Transplanting something big is not easy or quick. One needs the right equipment, and success only follows a complete preparation of the ground. In 1990 Orr began to prepare the soil for transplanting his educational model with a talk to the college titled "What Good Is a Great College if You Don't Have a Decent Planet to Put It On?" This essay remains powerfully relevant. The heart of its subject is found in a simple declarative statement, "First, we should recognize that all education is environmental education," and in an incisive question, "Does four years at Oberlin make our graduates better planetary citizens or does it make them, in Wendell Berry's words, 'itinerant professional vandals'?" At that time, however, there were few people ready to act upon Orr's advice: accept that it is highly educated people who have ravaged Earth and revise our educational enterprise to turn out ecologically literate graduates who have an environmentally focused agenda.

When Orr arrived at Oberlin, the Environmental Studies Program had a half-time secretary and a basement office. At that time, the program was graduating fewer that ten students each year. Orr was the only faculty member exclusively dedicated to the program, and a committee of faculty from several other departments oversaw environmental studies. In Orr's first meetings with the committee, he asked its members for their vision of the program. Unlike other disciplines, environmental studies lacked a physical presence on campus; and the committee felt that an environmental center was a priority. This charge from the committee resonated with Orr's understanding that physical connections are required for effective environmental education.

Orr took the lead in creating the center; however, he was soon frustrated. He put it this way: "In academic institutions the clock speed, the sense of urgency, is incredibly different from that of other groups in the not-for-profit sector. You'll have a committee meeting, then another and another. You know, people sit around and talk ad nauseam." For a visionary such as Orr, the slow pace of academic decision making was demoralizing.

Beginning in 1992 Orr used one of his seminar courses as a forum to discuss what kind of physical presence would nurture the imagination and permit mind and hands to work together. In Orr's words, "The design, construction, and operation of academic buildings [and grounds] can be a liberal education in a microcosm that includes virtually every discipline in the catalog." A parade of the best architects and designers in the United States contributed ideas for such a building to Orr's seminar class, and the brainstorming for the environmental studies facility was under way.

Orr told the Oberlin administration that he would raise all the money for the environmental center by himself. Several years passed. Meetings, memos, and words filled the time, but no decision was made to authorize the building or allow him to raise funds. As with the situation at North Carolina two decades earlier, some people at Oberlin may have wished that Orr would stick to his teaching or just go away. Fortunately, however, he got the green light. To build the center, Orr was allowed to raise money from whatever sources he could find, except contributors or potential contributors to Oberlin. And the development office would not help him. Nonetheless, the money started rolling in, and after fifteen months he received the naming gift—three million dollars from Adam Joseph Lewis and his family.

As soon as approval was given for the environmental studies center, Orr invited students, townspeople, faculty, staff, and all who were interested to participate in public-design meetings and other planning venues. In spring 1996, after acquiring the naming gift, Oberlin designated William McDonough architect of the Lewis Center. On September 25, 1998, an earthquake registering 5.2 on the Richter scale shook northern Ohio, a rare occurrence. About one hour later, ground was broken for the center. The road had not always been smooth, but Orr had held course, and, on September 15, 2000, the Adam Joseph Lewis Center was formally dedicated.

The Lewis Center is a two-story, 13,600-square-foot building equipped with rooftop photovoltaic panels for generating electricity from sunlight that in its first years of operation provided over half of the building's energy. Building-efficiency improvements have reduced energy use, and with the addition of a second array of photovoltaic panels next to the building in fall 2004, all of the center's energy will be acquired on-site. The Lewis Center is a high-performance building that in its first years used about one-third the energy that a similar building of standard construction in northern Ohio requires. How is this possible?

The sun provides energy to the building. Orientation, window placement, and internal design maximize passive solar heating and lighting. Roof and walls are constructed to reduce heat loss in winter and decrease heat gain in summer. Geothermal wells, in conjunction with heat pumps, maintain building temperature. Motion detectors and light sensors are integrated to provide energy-efficient lighting only when needed. Window design and overhanging eaves are some of the center's other energy-efficient features. The long-term goal is for the Lewis Center to be climatically neutral—designed to change as required to meet ultimately all of its energy needs with the net release of zero carbon dioxide.

The construction plan, however, called for more than just energy efficiency. The building process and materials were selected to minimize harm to human health and the environment in Oberlin and elsewhere, during construction and after. Recycled materials were used, and the wood construction materials were from forests certified as sustainably managed. Organic compounds that do not readily evaporate into the air were utilized throughout, and toxic glues, paints, and adhesives were kept to a minimum. In strict compliance with the commitment to nontoxic materials, the upholstery on the auditorium seats can be composted for the garden or eaten directly with no ill effects. The carpet and raised floor are leased, and the carpet will be replaced as required and recycled into new carpet. Mimicking natural wetlands, a "living machine" stocked with a host of microbes, plants, and animals, powered by sunlight, purifies all waste water. The reclaimed water is recycled for flushing toilets and will be used for watering the orchard, garden, and other plantings around the building once health department approval is granted.

Oberlin College is a place where people come to learn; thus the paramount element in the Lewis Center project—from conception into operation—has been education. Years of planning, design, and construction enabled a large number of people, on and off campus, to participate in this most unusual educational experience. The living machine, operated by students and staff, is used for research and, just by its prominent physical presence, demonstrates how the biological world breaks down or doesn't break down our wastes. If someone pours enough paint, commercial cleaner, or any of many other common household items down a drain, the resilience and ultimately the effectiveness of the living machine to cleanse waste is compromised. Such short-term negative feedback teaches that our actions in the larger world do, indeed, have consequences.

The building is designed over its life to be efficient enough to be powered by sunlight and not to generate heat-trapping gas emissions. The National Renewable Energy Laboratory (NREL) has collaborated with Oberlin to analyze the building's energetics; however, it will be a while before these goals can be shown to be achievable. In spring 2001 data on light levels throughout the building over a several day period established the Lewis Center as among the best, if not the best, naturally lighted building ever assessed by NREL. Data on items such as energy use, output from photovoltaic panels, temperature, air flow, and air-quality parameters are monitored and analyzed by students working with environmental studies faculty member John Petersen. The center is an "organism" with a host of sophisticated sensors built into it and placed around it, so that the building's users can see clearly its relations with the larger environment, which are monitored by the center's weather station. As this evaluation process proceeds, students and others will see how the building actually performs environmentally and can compare it with other buildings. Adjustments can be made, and innovations employed to make the Lewis Center's environmental impact as benign as possible. And it should be noted, when a full accounting is completed, it is anticipated that per square foot the basic, no-frills cost for building the Lewis Center will be equal to or less than that of standard construction.

An orchard, garden, and a miniature native-Ohio forest grow around the building, providing opportunities for students to see, understand, and interact with the plants around them. Next to the Lewis Center a pond and drainage area create habitat for small animals and fish, as well as reeds, grasses, pond lilies, and other native plants. Almost everything about the center allows for doing with the hands while learning with the mind, reflecting the fundamental lesson about education that Orr learned at Meadowcreek.

In response to the success of the center and its programs, other large gifts have followed for the purposes of renovating a neighboring building to be an environmental science research and teaching laboratory, endowing a chair in environmental studies (now held by Katy Janda), and installing a second photovoltaic array next to the Lewis Center.

Oberlin College now has one building, the Lewis Center, that is moving toward releasing no heat-trapping gases and thereby achieving climatic neutrality. With this ongoing project in hand, Orr proposed to the college, and all institutions of higher learning, that the new goal should be for the entire college to be climatically neutral by 2020. This is a noble and critical challenge since climate experts expect human activities are going to push the Earth's temperature up as much as 10°F in the next one hundred years unless we make radical reductions in our use of fossil fuels and in other activities that increase heat-trapping gases.

Orr proposed the 2020 goal of climatic neutrality to the Oberlin community in late 1999, and by March 2000 a formal proposal to do a feasibility study of Oberlin was presented to the administration. The Educational Foundation of America funded the feasibility study of achieving this goal at Oberlin. The final report from the study, conducted by the Rocky Mountain Institute, was presented to Nancy Dye, Oberlin's president, in January 2002. Oberlin's policy on climate change is now part of a comprehensive environmental policy approved by the Oberlin Board of Trustees in March 2004. Although the policy is approved, as with the Lewis Center, it will take a formidable effort to have the Oberlin community—students, trustees, administrators, alumni, faculty, and staff—decide upon and then take the actions required to make the college climatically neutral.

Orr's vision connects campuses to their environs, human and natural. In 1993, he learned of a course being designed at Rensselaer Polytechnic Institute in Troy, New York, that used the local Hudson River as a microcosm of the world to teach students about place and introduce them to the challenges of creating sustainable patterns of living. He realized the value of this approach and immediately organized and taught in 1994 the course "Introduction to the Black River Watershed." This was the springboard for the Black River Environmental Education Partnership Project that brings together Oberlin students, local school teachers, and other interested community members. Outreach programs in kindergarten through high school, as well as community-based projects, not only engender appreciation for and connection with the local environment but also help clean up the watershed.

From its beginning the Environmental Studies Program worked with the Oberlin Student Cooperative Association (OSCA) to support locally and organically grown food, and the effort intensified with a food audit at Oberlin in 1988 organized by Orr while he was still at Meadowcreek. In 1996 Orr and others supported the creation of the Oberlin Sustainable Agriculture Project that worked with OSCA to establish a community-supported farm. As a response to urban sprawl spreading into the farming communities around Oberlin, the Ecological Design Innovation Center, another Orr-catalyzed organization headed by environmental studies graduate Brad Masi, was created on a 70-acre landholding of the college. This group quickly became an independent not-for-profit organization and is a community center for restoration ecology, organic and sustainable farming, land-use planning, water conservation, and other outreach educational initiatives.

In 1998 the Environmental Studies Program initiated a speaker series that led to the formation of the Cleveland Green Building

Coalition, which would apply many of the ecological design principles used in the Lewis Center and elsewhere to redevelopment projects in Cleveland. Under the leadership of Sadhu Johnston, a 1998 Oberlin environmental studies graduate, this coalition, as its first project, raised about three million dollars to purchase and renovate a downtown Cleveland bank building in compliance with standards established by the U.S. Green Building Council's Leadership in Energy and Environmental Design (LEED) rating system. In May 2000 eight Oberlin students completed a proposal to address student housing needs that would embody "green" design and help reinvigorate Oberlin's downtown area. Another Environmental Studies Program graduate, Naomi Sabel, along with two other classmates, is putting together a multimillion-dollar project to revitalize a city block in downtown Oberlin, which will employ ecological design principles used in the Lewis Center. In 2003 Joseph Waltzer, who started the Black River Cafe where Orr often conducts business over a locally grown organic meal, started a second "green" restaurant in Oberlin. Another environmental studies student, Sara Kotok, established Kotok's Market to provide local produce to the city on a regular basis. These examples illustrate what Orr calls the "multiplier effect." This ever-growing cadre of environmentally educated students—Oberlin now graduates more than fifty environmental studies majors each year—goes out and applies what they have envisioned and learned. They are not specialists, but they have a vision and have learned to think in terms of systems, work with others, and realize their vision in a particular place and environment.

Orr has exerted influence well beyond Oberlin. The food study at Hendrix College in 1987 was the beginning of the green campus movement. Thousands of students now attend a plethora of environmental conferences across the United States every year and many have been inspired by Orr. He gives 40 to 50 presentations annually: in the first three months of 2004 he gave over 20 talks at colleges and universities, including Rice, Carnegie-Mellon, Warren Wilson, Yale, and Harvard, as well as many presentations for groups such as the Montessori

Association in San Diego, the Pittsburgh Parks Conservatory, and the Lakeside Association in Cleveland.

Although the clock speed of the academy has often frustrated Orr, he tells of remarkable progress: "The whole green campus movement began with that first food study at Hendrix in 1987 and at about the same time with April Smith's study at UCLA called 'Not in our backyard.' The result of this worldwide campus movement seventeen years later is the full campus environmental policy now approved by the Oberlin College trustees. We can see colleges as places connected to the biosphere with input-output studies. These connections are used as curricular materials and the studies themselves enable institutions to reduce their environmental impact."

Orr contends that the challenges before humanity in the twenty-first century call for universities and colleges to be the Lewis Center writ big. It is not the building but the philosophy behind it that is important. Orr put it this way, "Thoreau went to Walden because he wanted to drive some of the problems of living into a corner to examine them. In the Lewis Center we have driven a lot of the problems of sustainability into an acre and a quarter where they can be infused into the curriculum and into research projects: ecological restoration, ecological design, horticulture, ecological engineering, renewable energy technologies, data analysis and display. All of those things are a part of a laboratory in sustainability. Students are doing the calculations and gaining the analytic tools necessary to make a sustainable world. The Lewis Center functions as a system that is inherently educational." Meadowcreek proved to Orr that effective education and living well in a place are not just inseparable but compatible and necessary. He believes that colleges and universities are ethically bound to educate students to live well in their places by example and as part of the curriculum.

When Orr applied emerging ecological design principles to the creation of the Lewis Center, he "thought that within a couple of years of completion it would be superseded by other and better buildings. But it hasn't been, even after almost six years of operation. Not at least in the United States. Rather, it is now the benchmark for academic architecture. More than a hundred places like Yale and Stanford are replicating the building or aspects of it, but no one is trying to do a next generation Lewis Center. Why?" Orr suspects that the philosophy behind the Lewis Center requires too radical a shift from our present one. We are still trying to make the natural world fit us, while the Lewis Center attempts the opposite: we need to adopt ecological principles to be compatible with 3.8 billion years of evolution.

Orr expresses his ideas and thinking about environmental education in three books: *Ecological Literacy: Education and the Transition to a Postmodern World; Earth in Mind: On Education, Environment, and the Human Prospect;* and *The Nature of Design: Ecology, Culture, and Human Intention.* In these writings he argues and clearly outlines everything he has done since coming to Oberlin in 1990. However, Orr did not discover the intellectual landscape upon which he built his educational alternatives—this had been discovered, mapped, cleared, and plowed by John Muir, Aldo Leopold, Gene Odum, Rachel Carson, René Dubos, Ian McHarg, Paul Ehrlich, E. F. Schumacher, Herman Daly, Peter Raven, Edward O. Wilson, and many more. Instead, Orr's contribution to education is similar to that of Wes Jackson's to agriculture and Amory Lovins's to energy. Orr connects what has been happening in the world to how and what we teach in the industrialized world. He places Western culture's incomplete and flawed educational agenda in the political arena and asks educational institutions not only to accept biophysical reality but also to base education upon this reality. This task is daunting because what these environmental visionaries have discovered—and Orr now advocates—is still virgin territory or, even worse, terra incognita for many in the Academy and for most of the general public.

Orr is a generalist connecting information and perspectives from a variety of fields. In his creative scholarship he works with isolated gems that others have ferreted out and makes of them a practical whole. His holistic program blends synthetic scholarship with practical action, which is not only unusual but also much needed in the Academy. For Orr, the enduring value of such creative scholarship is in its application toward alleviating injustice, inequality, and societal disintegration. He knows that if humanity is unable to embrace an ecological perspective before life support slips away, Western culture's hard-won freedoms—religious, political, and economic—which have emancipated humanity, may be lost; even now these freedoms are tempered by the biophysical limits confronting us. As an academic, Orr believes deeply that creative scholarship must be used to address the overwhelming odds against the survival of our civilization.

Orr's influence has spread through his writings. *Earth in Mind* has resonated with many in higher education, and even after ten years it is in demand—400 copies sold in December 2003. All first-year students at several schools have been required to read it. Architecture and business schools have made it core reading, and members of a faculty committee preparing grants in ecological literacy were all given copies. In recognition of the book's success and importance, its publisher has issued a special tenth-anniversary edition in 2004.

Throughout the United States and elsewhere, the many initiatives in environmental education are harbingers of the age of ecology. At the same time they are like germinated seedlings of a new tree species in a forest of well-established giants—vulnerable and fragile in a habitat not conducive to their survival.

We in education, who not only know the data but also believe the disastrous consequences they foretell, have an immensely challenging task before us. As Orr told me on Oberlin graduation day 2000: "The Academy, as you know it, is badly structured to turn out people who care about or are competent to preserve the creation. That's why in *Earth in Mind* I mention—on days like this, when the kids graduate— you can hear, if you are real quiet, the earth just groan and sigh because these kids on balance will be more destructive than they will be constructive in an ecological sense. We wallow in self-congratulation. Look at us humans: we act as if we are the pinnacle of everything we do, which is so important that it overwhelms the ecosphere. This self-importance pervades the Academy, that incredible sense that somehow we are different, without limits. This self-importance shows in our achievements, in our institutions, in our history. Where the hell does evolution fit in? Where is the sense of vulnerability in a species that is very far out on a fragile limb?"

Over recent centuries the big ideas of religious, political, and economic freedom combined with the manipulative capacity realized through the natural sciences to emancipate Western civilization. This culture not only civilized the world but also spawned the belief that humanity is exceptional and exempt from what constrains the rest of life. Estranged from the living world and cocooned in a human-mediated existence of technological achievements, we believe and thus teach that limits do not apply to us. "Throughout history," Orr writes, "humans have steadily triumphed over all of those things that managed us: myth, superstition, religion, taboo, and above all, technological incompetence. Our task now is to replace these constraints with some combination of law, culture, and a rekindled reverence for all life." We have escaped from that which constrained past caravans of humanity, but we are incompetent at managing Earth. We shall have to manage ourselves, or the laws of nature will do it for us and curtail our excesses.

What role does education have in this management? Perhaps no greater educational challenge exists than emotionally reassociating humanity with the biosphere. In industrialized civilization we venture from our technological cocoons to sample the living world. Our culturally inculcated perspective allows us to be awed, to be filled with wonder, but, by and large, not to connect with the rest of life.

How can this emotional bond be facilitated? If we look at the childhood experiences of environmental leaders and scientists who champion the preservation of the biological world, a common theme is an intimate association with a non-human-mediated natural world. Edward O. Wilson, the eminent biologist and a world-renowned advocate for biodiversity preservation, proposes that these experiences, especially when they occur in childhood, nurture what he calls biophilia, a love of other living things.

What if every class in every educational institution, from preschool through university, adopted a local ecosystem as a means of informing its educational program? The students would observe, characterize, and respect the ecosystem, which would be the starting place for all aspects of their education. The ecosystem perspective thus acquired would enable us to assess our relations with the world. Then perhaps a shift to communal kindredness with all life might be cultivated. Our intimate association with other life would then inform our education and all that followed, rather than the occasional meaningful encounter with the biotic world marginally influencing how we live. This, I believe, is what Meadowcreek deeply imprinted into David Orr's very being and what he is trying to have all of us understand.

TEN

Can We Change, Will We Change?

> [P]eople fail to adapt for at least three reasons: first, they may not
> see, understand or fully grasp the significance of the threat before
> them, and therefore cannot see its ramifications; second, the chal-
> lenge may simply be too great for the culture's adaptive capacity,
> something that cannot be known until all options have been tried
> and failed; and third, the anxiety, pain or conflict involved may be
> too stressful.
>
> JO ANNE VAN TILBURG, *Easter Island*

After almost a century of research and discussion, scientists have
reached consensus: our present patterns of living cannot persist without serious negative consequences, and some of the damage already
done will take millennia to reverse, if it can be at all. Humans have
become a dominant evolutionary force on the planet, radically changing the dynamic equilibrium that has made Earth a habitable, friendly
home for us. The uncomfortable truth is that the current scale and
character of human activities are decreasing the planet's life-support
capacity both in known and in unanticipated ways, and perhaps more
rapidly than we suspect.

Although science has established that humanity is headed into
ever-more treacherous waters at full sail, serious political discussions in
the halls of power, and within the general public, have hardly begun.
After noteworthy leadership in the United States that produced a con-
sensus-based avalanche of environmental legislation in the mid 1960s
and into the next decade, population and other environmental concerns then became contentious territory in the 1980s. The national and
international momentum created by the visionary recommendations
of the Rockefeller Commission on Population and the American
Future in 1972 dissipated over the next decade. With the Reagan
administration's Mexico City policy in 1984 that discarded, among

other things, comprehensive family planning as a primary means for arresting population growth, the U.S. government abandoned the wisdom of that commission's recommendations. In 1973 the Endangered Species Act passed 390 to 12 in the House of Representatives and 92 to 0 in the Senate. However, as Dave Foreman's story highlights, the 1980s witnessed the aggressive promotion of road building for oil and other resource extraction in roadless areas of national forests and in other wilderness areas that harbored some of the very species and ecosystems the Endangered Species Act was intended to protect.

By the 1990s the ecological perspective had been marginalized to the point that most politicians not only avoided discussing the negative aspects of both human overpopulation and overconsumption, but also kept a low profile on all environmental issues. A few years earlier, Al Gore's 1988 presidential campaign emphasized his concern about ozone depletion and global warming, but he was ridiculed and chided for such positions. As Clinton's vice presidential running mate in 1992, Gore continued to speak up for environmental concerns. After the election, however, Gore's lack of support for Terri Swearingen's campaign against the Waste Technologies Industries (WTI) incinerator indicates that strong environmental positions were by then perceived as a severe political liability in the United States, especially if they conflicted with established economic interests. Nonetheless, Secretary of the Interior Bruce Babbitt, Vice President Al Gore, and others in the Clinton administration and Congress did prevent the anti-environmental agenda of the mid 1990s, led by Newt Gingrich, from being fully implemented. And, as a result of several decades of scientific research and in response to over 2 million public comments, in its final days the Clinton team did set aside 58 million acres of national forest to remain roadless and undeveloped. However, this land preservation was achieved without congressional consensus, and it was mostly an executive branch, partisan accomplishment.

By the beginning of the twenty-first century the continued trend of indifference in the United States toward the environment has only intensified. Internationally the United States abandoned the Kyoto Protocol, thereby substantially reducing the effectiveness of the first attempt among the world's nations to lower dangerous carbon dioxide emissions worldwide. At home the proposed Energy Policy Act of 2003 and other pro-industry legislation have weakened the Clean Air Act and opened yet more wilderness areas to logging, mining, and drilling. When George W. Bush gave his State of the Union address in January 2002, he mentioned the environment in a single sentence referring to a "cleaner environment." Likewise, in the spring of the same year a talk by senior Democratic Senator Ted Kennedy at the National Press Club covered just about everything, but the word "environment" was not even mentioned. In addition, the 2004 State of the Union addresses given by top Republican and Democratic leaders—George Bush, Nancy Pelosi, and Tom Daschle—not only failed to include any environmental issue but also neglected to mention the word "environment." Although politicians—including Bush, Daschle, Kennedy, and Pelosi—do mention environmental issues sometimes, their discussion of such issues is sporadic and not central to their political agendas. The nascent, nonpartisan political approach to environmental issues of the early 1970s has not matured into a consistent or dominant pro-environment national policy.

Over the past two decades the United States, for the most part, abandoned its world leadership role in population and environmental issues, with the approval of the electorate. In contrast, over the same time, information on and general awareness of environmental concerns have become almost universal among the citizens of the United States and of much of the rest of the world, developed and developing. This contradiction was brought home to me when I was a Population Institute delegate at the United Nations 2002 World Summit on Sustainable Development (WSSD) in Johannesburg, South Africa.

Despite the general awareness in the United States and throughout the world of the importance of this second world conference on sustainable development, the Bush administration and much of the leadership in the United States treated it as an event that did not merit their full attention although delegates had been sent. The conference focused on a wide range of objectives for people everywhere: providing food, clean water, sanitation, health care, education, jobs, a pollution-free environment, and transportation; eliminating corruption, war, violence, exploitation of women and children; and using renewable and nonrenewable resources in sustainable ways—all important goals that most people in the United States support. However, at the WSSD the U.S. government representatives appeared to have as one of their primary objectives the avoidance or elimination of direct consideration of population and consumption. These issues, however, underlie essentially all the others considered at the conference and in this book; the reduction of the world's population and of consumption of resources is a critical step in resolving all other environmental and social problems.

On the ground in Johannesburg, one minor and poorly attended talk highlighted for me the disjunction between our global environmental reality and the U.S. government agenda on environmental policy. Here, several speakers presented findings from the book *Stumbling toward Sustainable Development*, in which forty-two accomplished professionals and scholars had measured the United States' movement toward sustainable development since the 1992 UN conference in Rio de Janeiro, the first world conference on sustainable development. The concluding talk was given by Christopher Shays, a Republican member of the U.S. House of Representatives from Connecticut. He began by holding up a hand-drawn graph. The vertical axis showed the number of elected politicians and the horizontal axis gave their environmental commitment. All of the dots clustered between 0 and 20 percent commitment. He said that Democrats had a better environmental record than Republicans but that does not mean much; overall Democrats won't fight for strong environmental laws and Republicans do not support environmental legislation. He praised Al Gore for getting it basically right in his book *Earth in the Balance*. He contended that Gore could have been a great leader like Winston Churchill, but unlike Churchill, when the people weren't with him, Gore gave up his core principles. Shays then discussed his own perspectives on various environmental concerns, ending with, "Nothing will happen until the population wakes up with my daughter's generation that is learning in school what is happening. Unless we change, we will not have a livable world. There is no question about it."

Those of us who have dedicated our lives to teaching concur with Shays' belief that education is critical in order for humanity to recognize and act on the present threat to the environment, but educating children alone will not resolve the problems we face. My late father-in-law, the Reverend Donald Frazier, summarized it this way in a sermon he gave that reflected upon his sixty years in the ministry: "Of the many new congregations I observed, almost all of those that dedicated themselves first to educate religiously the children went on to fail. It is the adults who need their faith to be stirred." With this observation Frazier affirmed that children mimic adults; to maintain a livable world, young and old will have to change their perspectives together.

The task before humanity is nothing less than a total change from the dominant economic worldview to an ecological one. The difficulty in effecting this transition was made clear to me by what I saw on South African television several days after Shays' talk. Bill Moyers, the commentator and documentary filmmaker, had organized and was hosting in Johannesburg a special "summit" gathering of world-renowned people to consider sustainable development. Among the highlighted participants I heard and watched was Bjørn Lomborg, a Danish statistics instructor who had published in 2001 a thoroughly discredited book, *The Skeptical Environmentalist: Measuring the Real State of the World.*

Unlike so many other environmental books—from *A Sand County Almanac* to *Silent Spring* to *A Green History of the World* to *The Future of Life,* which give sober warnings in their arguments for change—*The Skeptical Environmentalist* announces, "Don't worry. Everything is just fine and getting better." Lomborg concludes his book with:

We are actually leaving the world a better place than when we got it and this is the really fantastic point about the real state of the world: that mankind's lot has vastly improved in every significant measurable field and that it is likely to continue to do so. . . .

Thus, this is the very message of the book: children born today—in both the industrialized world and developing countries—will live longer and be healthier, they will get more food, a better education, a higher standard of living, more leisure time and far more possibilities—without the global environment being destroyed.

And that is a beautiful world.

Published by Cambridge University Press, a well-respected academic press, and containing 82 pages of 2,930 individual notes and some 2,000 references, *The Skeptical Environmentalist* looks as if it were a solid piece of scholarship. *The Economist,* the *New York Times,* and *Washington Post,* among others, published reviews that lavished praise on Lomborg's "truth" about the real state of the world. The *Washington Post's* review ended, "Bjørn Lomborg's good news about the environment is bad news for Green ideologues. His richly informative, lucid book is now the place from which environmental policy decisions must be argued. In fact, *The Skeptical Environmentalist* is the most significant work on the environment since the appearance of its polar opposite, Rachel Carson's *Silent Spring,* in 1962. It is a magnificent achievement."

In *The Skeptical Environmentalist* Lomborg reflects the perspective of our economically centered culture. He heralds a universally beneficial, continuous progress, despite overwhelming evidence to the contrary, and the book's optimistic message has been widely embraced. Every few years an unrealistically optimistic book like Lomborg's elicits such responses: in the 1990s there were, among others, Gregg Easterbrook's *Moment on the Earth* and Julian Simon's *The Ultimate Resource 2.* Usually the dynamic tension between the mainstream belief in the status quo and the evidence that makes that belief no longer tenable keeps such books from dominating environmental debates. Lomborg's *The Skeptical Environmentalist* tipped the balance, however. The initial response of many politicians, commentators, and others who welcomed this message was to accept with enthusiasm this affirmation of what they believed, or wanted to believe, and to question the veracity of those who contend substantive change is necessary to avert disaster. The deceptive message of this book is extremely dangerous, for as Winston Churchill noted: "A lie gets halfway around the world before the truth has a chance to get its pants on." Furthermore, a deception that confirms a widely held worldview is not only readily accepted but also hardens belief in that worldview.

Edward O. Wilson summed up the pernicious effects of the publication of *The Skeptical Environmentalist:* "My greatest regret about the Lomborg scam is the extraordinary amount of scientific talent that has to be expended to combat it in the media. We will always have contrarians like Lomborg whose sallies are characterized by willful ignorance, selective quotations, disregard for communication with genuine experts, and destructive campaigning to attract the attention of the media rather than scientists. They are the parasite load on scholars who earn success through the slow process of peer review and approval." In other words, the scientific community was diverted from doing important science to chase the lie.

In contrast to the enthusiastic reception of Lomborg's book in the popular media, the two most prestigious peer-reviewed science journals in the world, *Nature* and *Science,* published scathing reviews, as did *Scientific American,* the well-respected general-audience science magazine. The Union of Concerned Scientists, an organization of scientists dedicated to giving the general public science-based policy options, and *Grist Magazine,* an electronic environmental publication, together provided twelve pieces explaining the manifold distortions and fallacies in Lomborg's pronouncements. These articles conclude that Lomborg presented nothing new or controversial concerning the many positive trends he highlighted, but the contention that the current state of the world is good for humanity and will certainly get even better is unfounded. The innumerable and ever-growing list of toxins released into the environment, the present and future extinctions of species and ecosystems, and the climate change upon us that is certain to accelerate—all indicate monumental disruptions now and ahead. To project the successes of the past onto the future, as Lomborg did, is surely wrong.

Unfortunately, the world's political leadership and a majority of the general public are receptive to messages like Lomborg's, because they reassure us that fundamental change in our behavior is unnecessary. Many readers took Lomborg at face value because of his ability to manipulate and misrepresent data; in addition, their belief in the durability of their culturally advocated lifestyles were affirmed by the half-truths in his book. Although Lomborg apparently considers his book valid, clearly he understands little about the biological and physical world he attempts to describe.

Although Lomborg's impact may fade now that *The Skeptical Environmentalist* has been discredited by a host of eminent scientists, the book's immediate appeal shows how deeply Western culture holds the belief that the neoclassical, "free" market worldview is the linchpin for human progress. Furthermore, the controversy surrounding the book demonstrates that the prevalent economic worldview is not easily debunked by empirical knowledge of biological and physical reality. The phenomenal scientific, technological, and social achievements of this economic arrangement are undeniable and are what Lomborg finds affirmative in the past and projects into the future. His thesis that past accomplishments guarantee future success resonates with those who believe our economic system is responsible for humanity's progress and can grow without limits to deliver the good life to everyone.

The people described and the stories told here warn humanity, however, that it is the present and future consequences of our past successes that are crucial—most people fail to grasp this. Ecological foot-printing and other analytic tools that approximate Earth's carrying capacity for humans provide evidence that the scale and character of human activities are eating deeply into natural resources and thus diminishing the planet's present and future life support. We are liquidating Earth's ecological wealth without knowing how much we need to sustain us at the level to which we are accustomed or how many of us may be sustained at this level or for how long. Most people do not believe, or are not aware of, the scientific consensus that the current level and character of human activity is most likely unsustainable. The scientific consensus on the state of the world is that global civilization is unquestionably in jeopardy, yet most people are either ignorant of this conclusion or are unable to accept it because it conflicts with deeply held beliefs. And if humanity is to tip the balance toward survival, the cultural change required will have to be accomplished quickly—in years and decades, not centuries.

CAN WE CHANGE?

Successful cultures have acquired their character and resilience in innumerable field trials mastered and handed down by their ancestors. Change comes hard, often under circumstances in which tragedy hangs in the balance. Failure is visited upon those cultures that do not recognize or believe change is necessary, that do not have the capacity to change, or that do not have the will to change. Paleoarcheological and historical analyses provide evidence of countless local failures across the globe catalyzed in large part by impoverished natural resources, among them: many Polynesian catastrophes in the Pacific, including Rapa Nui (Easter Island), Mangaia, Mangareva, Henderson, and Pitcairn; in the Americas, Mayan and Anasazi cultures; in the Mediterranean, the Maltese islands; in the Middle East, Sumer, Egypt, Mycenae, Petra, and others; and in India's Indus valley, Harappan culture. The tragedy humanity now courts gives every indication of being far more serious and widespread than past local failures.

The profusion of human invention that has engulfed Earth was catalyzed by the Enlightenment some four centuries ago. It set on course a culture that emancipated itself from the tethers of confining religions, restrictive economies, and repressive governments. Religious freedom, open economies, and democratic patterns of governance blended with humanity's expansionary nature to create a culture that envisioned, and championed, a seductive belief of no limits, at a time when Earth appeared vast and without limits. In radical contrast to the past in which these great ideas found expression, humanity now faces a totally different vista. Today, in fact, we are constrained by a finite Earth overflowing with humans and their artifacts. A successful cultural shift requires that we accept the limits of Earth and its resources, an idea that contradicts that which delivered the present bounty to so many of us. We require a belief system that embraces with joy and celebration humanity's limitless capacity for spiritual and intellectual growth while at the same time accepting and appreciating the physical laws and limits governing the biosphere. Can we shift from an economically centered pattern of living to adopt lifestyles compatible with physical and biological limits?

In making a stab at answering this question, we must begin by looking at human behavior and its relation to our biological success. In common with animals that sexually reproduce, we engage in many behaviors associated with acquiring mates and reproduction. We exhibit hierarchical relations within and among our social assemblages, including those between males and females, and we have a tendency toward violent behavior, especially males. Like some animals we are territorial both as individuals and social groups, and form tribal bonds beyond close kin. Humans constantly seek, when possible, to bring more resources under the control of their group. These fundamental behaviors are shared with numerous other social animals, vertebrates and invertebrates.

Humanity's spectacular evolutionary success, however, relates to several traits that are perhaps unique, or at least extremely well evolved in us in comparison to other animals. We are highly intelligent, use symbolic language for communication, excel at culture, and make long-term contracts with nonkin, thereby entering into cooperative arrangements beyond family and tribe. These traits have combined to endow human social groups with exceptionally complex and efficient relations that have permitted them to become exquisite exploiters of resources. As a consequence *Homo sapiens* has overwhelmed all other vertebrate competitors including several other *Homo* species. Among large organisms humans are the ultimate competitor—no other species can withstand the selective pressures we bring to bear—and the ultimate invasive species; we can live virtually anywhere and can dominate any landscape. Despite our adaptability and dominance, however, groups of humans have, more often than not, radically impoverished their habitats to the point of jeopardizing their survival.

Evolution and culture have teamed up to deliver to humanity the ultimate challenge: can a global, scientific, technological, economic culture impregnated with a belief in a boundless future shift to a belief in a constrained future, which will foster behaviors compatible with the biological and physical limits of a finite world? Can human beings, hardwired to exploit available resources, make such a shift? Edward O. Wilson explains the dilemma this way:

> The human brain evidently evolved to commit itself emotionally only to a small piece of geography, a limited band of kinsmen, and two or three generations into the future. To look neither far ahead nor far afield is elemental in a Darwinian sense. We are innately inclined to ignore any distant possibility not yet requiring examination. . . . Why do [humans] think in this short-sighted way? The reason is simple: it is a hardwired part of our Paleolithic heritage. For hundreds of millennia those who worked for short-term gain within a small circle of relatives and friends lived longer and left more offspring—even when their collective striving caused their chiefdoms and empires to crumble around them.

We may be hardwired to think shortsightedly, but there is evidence that we can also think about the future and plan for it when survival calls for such behavior. Although many human cultures have overexploited the biodiversity that provided them with life support and subsequently failed, other cultures have perceived limits in the resources upon which they depended and lived within those limits. Detailed studies of traditional hunter-gatherer societies such as the Kalahari !Kung in Africa establish that some hunter-gatherers live well within the constraints of the environments they inhabit. On an 8-square-mile island in the central Pacific, Nauruan culture existed for perhaps 3,000 years in virtual isolation without impoverishing biodiversity to the point of social collapse. On their 2-square-mile Pacific island, Tikopians, after ravaging their biological resources for 800 years, stabilized their population, restored biological resources (except those species driven to extinction), and achieved a comfortable lifestyle within the biological and physical limits of their island. Helena Norberg-Hodge witnessed such a pattern of durable habitation when she walked the valleys and mountain passes of Ladakh almost thirty years ago. These few examples show that humans can accept the limits of their immediate environments and adopt steady-state economies of the type recommended by Herman Daly.

Many past agricultural cultures—among them, the Polynesian, Mediterranean, American, Middle Eastern, and Indian civilizations mentioned earlier—have consumed their natural resources to a point where life-support systems are degraded and cultural disintegration has followed. The traditional cultures of Kalahari !Kung, Ladakhis, Nauruans, and Tikopians are atypical, however, in their ability to change and thereby continue. Why were these peoples, especially the agriculturalist Ladakhis and Tikopians, an exception to the rule? What appears most relevant to cultural survival is the clear and sometimes rapid negative feedback a people receives when it overexploits its fragile natural resources. Survival requires not only a recognition of negative feedback signals—rarity or extinction of birds, scarcity of intertidal fish and shellfish, loss of forest, smaller harvests from cultivated fields—but also the cultural capacity to adopt new behaviors that arrest impoverishment before irreparable damage occurs. For isolated cultures no escape is possible; their people clearly depend on the immediate habitat and know it is all they have. When a culture is confined to an island or particular region with no opportunity to obtain resources from other places by trade or migration, the consequences of behaviors may be obvious sooner, and to more people, thus forcing corrective actions. Humans are phenomenally adaptive and, as these examples demonstrate, have the capacity, although often untapped, to accept and respect limits in order to survive.

WILL WE CHANGE?

A culture creates its present and therefore its future through the stories its people tell, the stories they believe, and the stories that underlie their actions. The more consistent a culture's core stories are with biological and physical reality, the more likely its people are to live in a way compatible with ecological rules and thereby persist.

Many of the stories that underlie contemporary global culture have been shown to be false or to preclude adaptive behavior. The time has come for core stories that will lead us to preserve a world rich with biological diversity and resilient when faced with the vagaries of environmental change—thereby providing an enduring future for humanity. To accomplish this, we need a wholly different understanding and a new agenda.

If we hope to beat the overwhelming odds against abandoning an *econo*centric worldview for an *eco*centric one, humanity will have to use all of its knowledge and innate capacities—foremost human intelligence. We have among us visionaries of change who have exerted their extraordinary abilities to this end. I have written about a few of these men and women, and their stories may inspire myriad others like you and me to bring about changes that can make all the difference for those who follow. My hope is that if we all fully grasp the stories that express our connection to the rest of life and our absolute dependence upon the bugs and weeds of the world—and we tell these stories—they will become part of humanity's sacred beliefs and lay the foundation for a future of continuous progress. Through the actions of each one of us, global culture can embrace an urgently needed ecologically centered pattern of living. With this transformation we will take our rightful place among the rest of nature and accept with grace and humility our relations to Earth's diversity of life.

Notes

ONE: INTRODUCTION

2   An environmental perspective on the history of human habitation is given in Clive Ponting, *A Green History of the World: The Rise and Fall of Great Civilizations* (New York: Penguin Books, 1991); and on twentieth-century history in J. R. McNeill, *Something New under the Sun: An Environmental History of the Twentieth-Century World* (New York: W. W. Norton, 2000). Biological perspectives on the present and future are provided in Stuart L. Pimm, *The World According to Pimm: A Scientist Audits the Earth* (New York: McGraw-Hill, 2001); and Edward O. Wilson, *The Future of Life* (New York: Knopf, 2002). A summary of recent environmental data and trends is given in Paul Harrison and Fred Pearce, *AAAS Atlas of Population and Environment* (Berkeley: University of California Press, 2000).

      The negative trends listed and many more are provided in Ponting, *Green History;* McNeill, *Something New;* Pimm, *The World;* Wilson, *Future of Life;* Harrison and Pearce, *AAAS Atlas of Population;* and in the following annual publications: Worldwatch Institute, *State of the World, 1984–2004* (New York: W. W. Norton, 1984–2004); and Worldwatch Institute, *Vital Signs, 1992–2004* (New York: W. W. Norton, 1992–2004).

3   A historic assessment of why the environmental changes of the twentieth century are not possible in the twenty-first century is given in McNeill, *Something New.* Based upon energy considerations, Richard Heinberg arrives at the same conclusion in *The Party's Over: Oil, War*

and the Fate of Industrial Societies (Gabriola Island, British Columbia: New Society Publishers, 2003). The case for complexity eventually leading to the collapse of human societies for a host of reasons, including impoverishment of life support, is presented by Joseph Tainter in *The Collapse of Complex Societies* (London: Cambridge University Press, 1988).

4   Many have assessed the historic consequences of human patterns of living and, in light of current trends, have advocated adopting an ecologically centered worldview, among them: Aldo Leopold, *A Sand County Almanac and Sketches Here and There* (Oxford: Oxford University Press, 1949); Rachel Carson, *Silent Spring* (Boston: Houghton Mifflin, 1962); E. F. Schumacher, *Small Is Beautiful: Economics as if People Mattered* (New York: Harper & Row, 1973); Wendell Berry, *The Unsettling of America: Culture and Agriculture* (San Francisco: Sierra Club Books, 1977); Thomas Berry, *The Dream of the Earth* (San Francisco: Sierra Club Books, 1988); Edward O. Wilson, *The Diversity of Life* (Cambridge: Harvard University Press, Belknap Press, 1992); Edward O. Wilson, *Consilience* (New York: Knopf, 1998); Thomas Berry, *The Great Work: Our Way into the Future* (New York: Bell Tower, 1999); Paul Hawken, Amory Lovins, and L. Hunter Lovins, *Natural Capitalism: Creating the Next Industrial Revolution* (Boston: Little, Brown, 1999); Carl N. McDaniel and John M. Gowdy, *Paradise for Sale: A Parable of Nature* (Berkeley: University of California Press, 2000); Lester R. Brown, *Eco-Economy: Building an Economy for the Earth* (New York: W. W. Norton, 2001); and Lester R. Brown, *Plan B: Rescuing a Planet under Stress and a Civilization in Trouble* (New York: W. W. Norton, 2003).

Jacques Barzun in *From Dawn to Decadence: Five Hundred Years of Western Cultural Life* (New York: HarperCollins, 2000) gives a thorough, if not comprehensive, presentation of the historic trends and facts, which elucidates the emancipations in religion, economics, and government that have enabled Western culture to be the domnant global civilizing force of the past several centuries.

TWO: CLEAN AIR

Terri Swearingen quotes not identified below are from my interview with her on March 12–15, 2001.

5   Epigraph: Terri Swearingen (presentation, Canvassers Conference, Ohio State University, July 22, 2000).

Major sources on the WTI incinerator controversy are: Terri Swearingen, "Chronology of Waste Technologies Industries (WTI) Violations and Other Relevant Information," (Tri-State Environmental Council, November 27, 1992), 1–61; Jim Schwab, "East Liverpool: Landfill in the Sky," in *Deeper Shades of GREEN: The Rise of Blue-Collar and Minority Environmentalism in America* (San Francisco: Sierra Club Books, 1994), 104–159; Benjamin Davy, "East Liverpool, Ohio: A Case Study in Incinerator Siting" in *Essential Injustice: When Legal Institutions Cannot Resolve Environmental and Land Use Disputes* (Vienna: Springer-Verlag, 1997), 89–122; Jake Tapper, "The Town That Haunts Al Gore," Salon.com, April 26, 2000, http://salon.com/politics2000/feature/2000/04/26/gore/print.html.

6   Swearingen quotation ("I'm a registered nurse, . . . I do have a few letters following my name.") is from Terri Swearingen, "Revolution from the Heart of Nature" (speech given at Bioneers 2000 conference, San Rafael, Calif., October 20–22, 2000).

Problems with WTI location and other information of the WTI controversy given in Schwab, *Deeper Shades of GREEN*, 104–159; and Davy, *Essential Injustice*, 89–122.

The closest house is only 320 feet from the WTI facility (Tapper, "The Town That Haunts Al Gore").

7   The lack of information on the effects of the products of incineration on human health in: "The Health Impacts of Incineration," parts 1 and 2 *Waste Not*, no. 276; no. 277 (Canton, N.Y.: Work on Waste USA, 1994).

A biography of Aldo Leopold is given in Marybeth Lorbiecki, *Aldo Leopold: A Fierce Green Fire* (Helena, Mont.: Falcon Publishing, 1996).

8   The story of Simon is given in Bill Heller, *A Good Day Has No Rain: The Truth about How Nuclear Test Fallout Contaminated Upstate New York* (Albany, N.Y.: Whitston Publishing, 2003). Total dose of 2.3 rads per person is in Heller, *A Good Day,* 46.

        The toxic insults we are inflicting on life were first documented for the public in Rachel Carson, *Silent Spring* (Boston: Houghton Mifflin, 1962).

10   Swearingen quotations ("I realized . . . stopped it," and "and I made up my mind . . . stop it?") in T. C. Brown, "Slow Burn," *Cleveland Plain Dealer,* Sunday magazine, February 1, 1998, pp. 14–17.

        Stories of Lois Gibbs and Love Canal and of Anne Anderson and Woburn, Massachusetts, environmental controversies are given in Fred Setterberg and Lenny Shavelson, *Toxic Nation* (New York: John Wiley, 1993); Schwab, *Deeper Shades of GREEN;* Sherry Cable and Charles Cable, *Environmental Problems, Grassroots Solutions: The Politics of Grassroots Environmental Conflict* (New York: St. Martins Press, 1995); Jonathan Harr, *A Civil Action* (New York: Vintage Books, 1995); and Davy, *Essential Injustice.*

12   Broken gas line incident is in Swearingen, "Chronology of Waste Technologies Industries," 7.

13   The story of the Tri-State Environmental Council can be found in Schwab, *Deeper Shades of GREEN,* 112–113.

14   Details of January 17, 1991, EPA hearing including quotations are in Schwab, *Deeper Shades of GREEN,* 115–119.

15   Voinovich's position on WTI, various quotations, and the events up to the arrest of the ELO 33 are in Schwab, *Deeper Shades of GREEN,* 122–126, 132–134; and T. C. Brown, "Citizens' Arrest," *Cleveland Plain Dealer,* June 7, 1992, pp. 8–9, 16, 20.

17   Voinovich "wanted" notice provided by Terri Swearingen.

18   "He's a weenie" incident and "weenie" campaign are from T. C. Brown, "Incinerator Opponents Bring Concerns Home to Voinovich," *Cleveland Plain Dealer,* October 28, 1991, sec. C, p. 1; T. Brown, "Curbside 'Weenie Roast' to Protest Waste Burner," *Cleveland Plain Dealer,* November 21, 1991, sec. C, p. 3; and Swearingen, "Revolution."

Swearingen quotation ("Frankly, I don't relish . . . gonna get roasted.") is from Swearingen, "Revolution."

Voinovich quotations ("Enough is enough."; "One . . . the protests."; "It's a safe site.") are from Mary Beth Lane, "Protests Convince Governor: Asks Moratorium on Incinerators," *Cleveland Plain Dealer,* December 5, 1991, sec. C, pp. 1–2; Mike Rutledge, "WTI Could Be Different Story If Done Again," *Review* (Columbus), December 5, 1991, p. 14.

20   Arrests at the Ohio EPA and trial of ELO 33, including Kaufman quotation ("I don't, . . . becomes meaningless.") are in Schwab, *Deeper Shades of GREEN,* 136–147.

Judge Walter J. Skinner's role in Woburn case is in Harr, *A Civil Action.*

21   Vice-president-elect Gore's quotations ("I'll tell . . . for a change." and "Serious questions . . . been answered.") are in "The 'Environmental President' He Won't Be: When Polluters Challenge, Al Gore Backs Down," in Jim Hightower and Phillip Frazer, eds., *The Hightower Lowdown* 2, no. 10 (October 2000): 1–3.

22   Carl Sagan's dustjacket comment in Al Gore, *Earth in the Balance: Ecology and the Human Spirit* (Houghton Mifflin: Boston, 1992).

23   Gore's failure to keep his promise not to issue a test burn permit is given in L. J. Davis, "Where Are You Al?" *Mother Jones* November/December 1993, 45–49.

William Reilly quotation (". . . said the Vice President . . . before leaving office.") is in "National Ombudsman Investigation of The East Liverpool WTI Hazardous Waste Incinerator: In Re: on the Record Investigation of William Reilly," by telephone at EPA offices in Washington, D.C., October 31, 2000, http://www.greenpeace.org/wti/reilly-testimonytext.html.

Greenpeace suit to prevent test burn and the court ruling to allow the test burn but to prevent incinerator operations until final EPA approval is given in Schwab, *Deeper Shades of GREEN,* 150–157; Davy, *Essential Injustice,* 104–111; Hightower and Frazer "The 'Environmental President,'" 1–3.

Aldrich ruling ("may pose an . . . and the environment.") is from Davy, *Essential Injustice,* 108.

25   Account of civil disobedience at the White House given in Ronald A. Taylor, "Protesters Arrested in White House," *Washington Times,* Friday, March 19, 1993, sec. B, p. 3.

Quotations from Swearingen, Allison, and Spencer when they met President Clinton are in "WTI Opponent Asks Clinton for Meeting," *Wheeling News Register,* April 18, 1993, p. 1; Jay Brookes, "WTI Opponents Talk to Clinton—Well, Sort Of," *Morning Journal* (Columbiana County), April 18, 1993, pp. 1, 9.

US EPA puts freeze on new commercial hazardous-waste incinerators is in Schwab, *Deeper Shades of GREEN,* 158.

26   Account of the arrests at Swiss embassy are from Staff, "23 Incinerator Foes Arrested at Embassy Protest," *Morning Journal* (Columbiana County), July 15, 1993, pp. 1, 7.

Ruling of Sixth Circuit Court of Appeals in Cincinnati given in Davy, *Essential Injustice,* 117–119.

27   Accomplishments of the WTI protesters—among them, setting a new agenda for management of toxic waste—are given in: Anne Becher, *American Environmental Leaders,* vol. 2 (Santa Barbara, Calif.: ABC-Cio, 2001), 782–785; Terri Swearingen et al., Tri-State Environmental Council et al., letter to Ms. Vernice Miller, National Environmental Justice Advisory Council, March 7, 2000; and Brown, "Slow Burn," 14–17.

EPA siting criteria for hazardous waste facilities presented in Environmental Protection Agency, "Sensitive Environments and the Siting of Hazardous Waste Management Facilities," EPA530-K-97-0003, May 1997.

29   Details of US EPA risk assessment of WTI, health problems, and WTI incidents given in Brown, "Slow Burn," 14–17; Swearingen, "Revolution"; Terri Swearingen et al., Tri-State Environmental Council et al., letter to Albert Gore, Jr., January 6, 2000; and Tapper, "The Town That Haunts Al Gore."

WTI's East Liverpool, Ohio, incinerator history from 1977 to the present, http//www.ohiocitizen/campaigns/wti/wti.html.

Motel 6 negotiations are in Swearingen, "Revolution."

30   The EPA ombudsman's ruling on WTI given in Cheryl Johns, "WTI under Fire: Ombudsman Suggests Halting Operations at Facility for at Least Six Months, Another Test Burn," *Review Online*, http://www .ohiocitizen.org/campaigns/wti/ shutelr.html.

31   Examples of the challenges to and failures of our established institutions to effectively employ scientific data with regard to the environmental issue of toxic pollution are in Setterberg and Shavelson, *Toxic Nation;* Schwab, *Deeper Shades of GREEN;* Harr, *A Civil Action;* and Davy, *Essential Injustice.*

The failure of scientific and legal institutions to effectively resolve environmental issues is discussed in O. Houck, "Tales from a Troubled Marriage: Science and Law in Environmental Policy," *Science* 302 (2003): 1926–1929.

32   The story of seat-cover fabric is from William McDonough and Michael Braungart, "The NEXT Industrial Revolution," *Atlantic Monthly*, October 1998, pp. 82–90.

33   Ray Anderson quotation ("We look . . . degrading it.") is in R. C. Anderson, "Interface Sustainability Report, 1997" (LaGrange, Ga.: Interface, 1997).

Data on Interface Corporation are from Interface, Inc. "2002 Annual Report" (Atlanta: Interface, 2002).

THREE: RESTORING WILDLANDS

Dave Foreman quotes not identified below are from my interview with him on July 31, 2001 and August 1, 2001.

34   Epigraph: Dave Foreman, *Confessions of an Eco-Warrior* (New York: Harmony Books, 1991), 38.

Dave Foreman's ideas and perspectives are presented in: Dave Foreman, *Confessions;* Dave Foreman and Howie Wolke, *The Big Outside:*

A Description Inventory of the Big Wilderness Areas of the United States,
rev. ed. (New York: Harmony Books, 1992); Dave Foreman, *The Lobo
Outback Funeral Home* (Boulder, Colo.: University Press of Colorado,
2000); Dave Foreman, *Rewilding North America: A Conservation Vision
for the 21st Century* (Washington, D.C.: Island Press, 2004); and Dave
Foreman's essays and "Around the Campfire" pieces in *Wild Earth,*
beginning in the first issue in spring 1991, "Around the Campfire,"
Wild Earth 1 (1991): 2, and ending in winter 2003–2004, "The Rewilding Institute," *Wild Earth* 13 (2003): 2–3.

A major source for information on Earth First! and Dave Foreman
is Susan Zakin, *Coyotes and Town Dogs: Earth First! and the Environmental Movement* (New York: Viking, 1993).

The Edward Abbey quotation about the Pinacate ("the final test of
desert rathood.") is from Zakin, *Coyotes,* 124.

35   The story of the Pinacate trip and the formation of Earth First! is in
Zakin, *Coyotes,* 115–130.

Edward Abbey's Pinacate Peak register entry ("Clumb up from Tule
Tank . . . Edward Abbey.") is from Zakin, *Coyotes,* 134.

The controversy over RARE II is in Zakin, *Coyotes,* 95.

36   Dave Foreman's connection to Edward Abbey is in Zakin, *Coyotes,* 26.

Details of Dave Foreman's first paying job as an environmentalist
working for the Wilderness Society given in Zakin, *Coyotes,* 27.

38   Events associated with Celia Hunter's coming to and leaving the
Wilderness Society are in Zakin, *Coyotes,* 88–93.

39   Events on July 4, 1980, in Moab, Utah, in Foreman, *Confessions,* 16.

40   The flyer announcing the objectives of the Wyoming rendezvous ("To
reinvigorate, enthuse, . . . howl at the moon.") is from Zakin, *Coyotes,*
142.

Events in Babbitt's hardware store given in Zakin, *Coyotes,* 148.

Edward Abbey quotation ("Earth First! Free the Colorado!") is
from Zakin, *Coyotes,* 150.

41   Increase in Sierra Club membership given in Zakin, *Coyotes,* 200.

Story of roadblocks at Bald Mountain and suit of the U.S. Forest
Service over RARE II is given in Zakin, *Coyotes,* 68–69, 256–259.

42   Dave Foreman quotation ("When death . . . fight for it.") is from
     Zakin, *Coyotes,* 293–294.

43   The Wildlands Project is described in Tom Butler, ed., *Wild Earth* 10
     (2000): 1–113; Michael E. Soulé and John Terborgh, eds., *Continental
     Conservation: Scientific Foundations of Regional Reserve Networks*
     (Washington, D.C.: Island Press, 1999); and Foreman, *Rewilding North
     America.*

44   Wilson quotation ("The somber . . . of life forms.") is in Edward O.
     Wilson, *The Future of Life* (New York: Knopf, 2002), 102.

45   Details of the North American biota and its demise are given in Tim
     Flannery, *The Eternal Frontier: An Ecological History of North America
     and Its People* (New York: Atlantic Monthly Press, 2001), 155–229.
     Other topics in this chapter from Flannery, *Eternal Frontier,* are: the
     fate of mammoth, mastodon (155–157, 194–205); short-faced bear
     (159–160, 213–214); condor (162); a dozen incipient species of grizzly
     bears (236–237); cat species and *Teratornis incredibilis* with 5-meter
     wing span (161).

46   Details of the sea otters on the west coast of North America and of
     jaguars and pumas on Barro Colorado Island given in Edward O. Wil-
     son, *The Diversity of Life* (Cambridge: Harvard University Press, 1992),
     164–168.

48   Extinction of the great auk is in Bruce A. Stein, Lynn S. Kutner, and
     Jonathan S. Adams, eds., *Precious Heritage: The Status of Biodiversity in
     the United States* (Oxford: Oxford University Press, 2000), 114.

     Extinctions of passenger pigeon and other biota, as well as the state
     of biological impoverishment in the United States, are given in Stein,
     Kutner, and Adams, *Precious Heritage,* 101, 115, 323–334.

     The ecological associations concept and the percentage of species
     and associations vulnerable to extinction are discussed in Stein, Kutner,
     and Adams, *Precious Heritage.*

49   The fate of North American megafauna after European arrival given in
     Flannery, *Eternal Frontier,* 236–237.

51 Foreman quotation ("[Here] the plants . . . on the wing.") and a discussion of Sky Islands are found in Dave Foreman et al., "The Sky Islands Wildlands Network," *Wild Earth* 10 (2000): 11–16.

Details leading to the establishment of the Gila Wilderness in 1924 can be found in Marybeth Lorbiecki, *Aldo Leopold, A Fierce Green Fire, an Illustrated Biography* (Helena, Mont.: Falcon, 1996), 83–96.

Leopold quotation ("Only the mountain . . . such a view.") is from Aldo Leopold, *A Sand County Almanac and Sketches Here and There* (Oxford: Oxford University Press, 1949), 129–130.

52 Leopold's assessment of ecosystem health is in Aldo Leopold, "Conservationist in Mexico," *Wild Earth* 10 (2000): 31–42.

53 The focal species concept and the focal species for Sky Island Wildlands Network are discussed in Dave Foreman et al., "The Elements of a Wildlands Network Conservation Plan: An Example from the Sky Islands," *Wild Earth* 10 (2000): 17–30.

The Wildlands Project corridors given in "Wildlands Conservation Planning Efforts," *Wild Earth* 10 (2000): 84–85; and were presented at a Wildlands workshop (Rowe Center, Massachusetts, March 1–3, 2002).

54 Florida habitat restoration efforts given in Stein, Kutner, and Adams, *Precious Heritage,* 132, 138, 196–197.

The saving of the Pinhook Swamp is given in Charles C. Mann, "Filling in Florida's Gaps: Species Protection Done Right?" *Science* 269 (1995): 318–320.

55 Creating the next great national park in Maine is the project of RESTORE: The North Woods, 9 Union Street, Hallowell, Maine 04347.

Economic analysis of the new park is given in Thomas Michael Power, "The Economic Impact of the Proposed Maine Woods National Park and Preserve" (Hallowell, Me.: RESTORE: The North Woods, 2001), 1–110.

56 Wilson quotation ("Great dreams, . . . human spirit.") from Edward O. Wilson, "A Personal Brief for the Wildlands Project," *Wild Earth* 10 (2000): 1–2.

The June 1983 issue of *Earth First!* published an article by Dave Foreman proposing that "[w]e should have large wilderness preserves of all our biological communities" in a preserve system of 716 million acres comprised of 50 separate preserves (Dave Foreman, Howie Wolke, and Bart Koehler, "The EARTH FIRST! Wilderness Preserve System," *Wild Earth* 1 [1991]: 33–38). This vision is also given in Dave Foreman, "Dreaming Big Wilderness," *Wild Earth* 1 (1991): 10–13, which is also included in Foreman's *Confessions*, 177–192.

FOUR: HEALTHY FARMS

Wes Jackson quotes not identified below are from my interview with him on October 20–21, 2000.

57 Epigraph: Wes Jackson, *New Roots for Agriculture* (San Francisco: Friends of the Earth, 1980), 4.

Wes Jackson's ideas and perspectives are presented in his three books and other publications: *New Roots for Agriculture; Altars of Unhewn Stone: Science and the Earth* (San Francisco: North Point Press, 1987); *Becoming Native to This Place* (Lexington: University Press of Kentucky, 1994); *The Land Report* and other publications from The Land Institute, 2440 E. Water Well Road, Salina, Kans. 67401; and: www.landinstitute.org.

The mission statement of The Land Institute appears in all issues of *The Land Report.*

58 Industrial agriculture and its consequences are discussed in: Jackson, *New Roots for Agriculture;* Wes Jackson, Wendell Berry, and Bruce Colman, eds., *Meeting the Expectations of the Land* (San Francisco: North Point Press, 1984); William E. Riebsame, "The United States Great Plains," in *The Earth as Transformed by Human Action,* ed. B. L. Turner II et al. (Cambridge: Cambridge University Press, 1990), 561–575; Judith D. Soule and Jon K. Piper, *Farming in Nature's Image* (Washington, D.C.: Island Press, 1992); Alan Wild, *Soils and the Environment* (Cambridge: Cambridge University Press, 1993); Michael H. Glantz,

Drought Follows the Plow (Cambridge: Cambridge University Press, 1994); Gary Gardner, "Shrinking Fields: Cropland Loss in a World of Eight Billion," Worldwatch Paper 131 (Washington, D.C.: Worldwatch Institute, 1996); and Andrew Kimbrell, ed., *The Fatal Harvest Reader: The Tragedy of Industrial Agriculture* (Washington, D.C.: Island Press, 2002).

    Erosion and soil formation rates are in David Pimentel et al., "Environmental and Economic Costs of Soil Erosion and Conservation Benefits," *Science* 267 (1995): 1117–1123.

59   Rachel Carson, *Silent Spring* (Boston: Houghton Mifflin, 1962); Frank Graham, Jr., *Since "Silent Spring"* (Boston: Houghton Mifflin, 1970).

    Sioux Indian story from Evan Eisenberg, "Back to Eden," *Atlantic Monthly,* November 1989, 59.

60   Biographical information on Wes Jackson and the first years of The Land Institute and its accomplishments can be found in Evan Eisenberg, "Back to Eden," *Atlantic Monthly,* November 1989, 57–59, 62, 65–68, 70–71, 74, 76–78, 81, 84–86, 88–89; and in *The Land Report,* a regular publication of The Land Institute.

63   E. F. Schumacher, *Small Is Beautiful: Economics as if People Mattered* (New York: Harper & Row, 1973).

    Story of hybridizations that led to bread wheat in the Middle East is in J. Bronowski, *The Ascent of Man* (Boston: Little, Brown, 1973), 65–68.

64   An overview of agricultural history is given in Clive Ponting, *A Green History of the World: The Rise and Fall of Great Civilizations* (New York: Penguin Books, 1991).

65   A thorough discussion of human use of biological productivity is found in Stuart L. Pimm, *The World According to Pimm: A Scientist Audits the Earth* (New York: McGraw-Hill, 2001).

67   Jackson quotation ("I think . . . widely adopted.") is from Jackson, *New Roots for Agriculture,* 4.

68   The difficulty of creating Jackson's vision of a perennial polyculture is considered in Jackson, *New Roots for Agriculture;* and Soule and Piper, *Farming in Nature's Image.*

70   The requirements for creating an ecologically based agriculture come from many articles in various *Land Reports;* and Soule and Piper, *Farming in Nature's Image.*

71   A discussion of John Todd's work is given in Nancy Jack Todd and John Todd, *Principles of Ecological Design: From Eco-Cities to Living Machines* (Berkeley, Calif.: North Atlantic Books, 1994). Contact information is: John Todd, Ocean Arks International, 176 Battery Street, Suite 1, Burlington, Vt. 05401, www.oceanarks.org.

73   Jackson quotations ("any natural . . . good indefinitely." and "Fundamentalism of . . . improve the soil.") and the story of ecological staircase are in Wes Jackson, "We May Want to Look into That . . . ," *Land Report* 55/56 (1996): 47.

74   Complexity research at The Land Institute discussed in "Complexity," *Land Report* 49 (1994): 2–19.

        Pimm quotation ("there is . . . in the system.") in Stuart Pimm, "Species Shakedown," *Land Report* 49 (1994): 11–12.

75   Complexity research results from David Van Tassel, "Community Assembly and Perennial Polyculture: 1997 Natural Systems Agriculture Research," *Land Report* 60 (1998): 17–18.

76   Agriculture appropriations legislation quotation ("make an . . . research needs.") is from *Land Report* 55/56 (1996): 59.

77   Natural systems agriculture reviewed by Stuart L. Pimm, "In Search of Perennial Solutions," *Nature* 389 (1997): 126.

        Sunshine farm is discussed in: Marty Bender, "The Sunshine Farm Takes Shape," *Land Report* 51 (1994): 14–18; David Tepfer, "Cattle on the Prairie," *Land Report* 51 (1994): 18–21; Jeremy Plotkin, "Chickens on the Sunshine Farm," *Land Report* 51 (1994): 22–23; and Marty Bender, "Energy in Agriculture and Society: Insights from the Sunshine Farm," (Land Institute, March 28, 2001).

78   Jackson's "homecoming" is considered in Jackson, *Becoming Native.*

79   Ecological footprinting is discussed in: Mathis Wackernagel and
     William Rees, *Our Ecological Footprint: Reducing Human Impact on the
     Earth* (Gabriola Island, British Columbia: New Society, 1996); and S.
     Levin, ed., *Encyclopedia of Biodiversity* (San Diego, Calif.: Academic
     Press, 2001), s.v. "ecological footprint, concept of."

80   Matfield Green projects are discussed in: Caroline Mahon, "Matfield
     Green School Re-Opens," *Land Report* 49 (1994): 24–25; Sue Kidd,
     "People, Land, and Community in Education: A Program for Educa-
     tors on Teaching Students about Place," *Land Report* 59 (1997): 16–17.

82   Consideration of the current Western economic system, which is
     unable to get the boundaries right, is in: Edward O. Wilson, *Con-
     silience* (New York: Knopf, 1998); Carl N. McDaniel and John M.
     Gowdy, *Paradise for Sale: A Parable of Nature* (Berkeley: University of
     California Press, 2000); and Eric A. Davidson, *You Can't Eat GNP:
     Economics as if Ecology Mattered* (Cambridge, Mass.: Perseus, 2000).
     See also chapter 7 ("Living in a Finite World") of this book and its ref-
     erences.

         Jackson quotation ("[W]e of Western . . . way, ecology.") is from
     Jackson, *Becoming Native*, 116.

FIVE: LIVING LOCALLY

*Helena Norberg-Hodge quotes not identified below are from my interview with
her on May 21–23, 2001.*

83   Epigraph: Helena Norberg-Hodge, *Ancient Futures: Learning from
     Ladakh* (San Francisco: Sierra Club Books, 1992), 180.
         Helena Norberg-Hodge's ideas and perspectives are presented in
     her books: *Ancient Futures;* Helena Norberg-Hodge, Todd Merrifield,
     and Steven Gorelick, *Bringing the Food Economy Home: The Social, Eco-
     logical and Economic Benefits of Local Food* (San Francisco: International
     Society for Ecology and Culture, 2000); and Helena Norberg-Hodge,
     Peter Goering, and John Page, *From the Ground Up: Rethinking Indus-
     trial Agriculture,* 2d rev. ed. (London: Zed Books, 2001).

Discussions of our relations to bonobos and chimpanzees and our animal origins can be found in: Jared Diamond, *The Third Chimpanzee: The Evolution and Future of the Human Animal* (New York: HarperCollins, 1992); and Frans De Waal and Frans Lanting, *Bonobo: The Forgotten Ape* (Berkeley: University of California Press, 1997); and on human nature in: Edward O. Wilson, *On Human Nature* (Cambridge: Harvard University Press, 1978); Edward O. Wilson, *Consilience* (New York: Knopf, 1998); Bobbi S. Low, *Why Sex Matters: A Darwinian Look at Human Behavior* (Princeton, N.J.: Princeton University Press, 2000); and Steven Pinker, *The Blank Slate: The Modern Denial of Human Nature* (New York: Allen Lane, 2002).

84 The importance of language in making humans unique is considered in: Diamond, *The Third Chimpanzee;* and Reg Morrison, *The Spirit in the Gene: Humanity's Proud Illusion and the Laws of Nature* (Ithaca, N.Y.: Cornell University Press, 1999).

The argument that humans have become a major evolutionary force on the planet is given in S. R. Palumbi, "Humans as the World's Greatest Evolutionary Force," *Science* 239 (2001): 1786–1790.

The propensity of humans to learn language is discussed in Steven Pinker, *The Language Instinct* (New York: HarperCollins, 1994).

85 The extinction of languages is considered in W. Wayt Gibbs, "Saving Dying Languages," *Scientific American* 287 (2002): 78–85.

In Norberg-Hodge, *Ancient Futures,* Norberg-Hodge's history that has enabled her to uniquely appreciate the importance of change in Ladakh; the subject of globalization everywhere; and information on Ladakh.

87 Norberg-Hodge quotation ("that human beings . . . utopian dreams.") is from Norberg-Hodge, *Ancient Futures,* 2.

89 Norberg-Hodge quotation ("Using limited . . . out of little.") is from Norberg-Hodge, *Ancient Futures,* 25.

Norberg-Hodge quotation ("What's the . . . live together.") is from Norberg-Hodge, *Ancient Futures,* 46.

Norberg-Hodge quotation ("One of the . . . more harshly.'") is from Norberg-Hodge, *Ancient Futures,* 56.

90   Norberg-Hodge quotation ("By the middle . . . to each other.") is from Norberg-Hodge, *Ancient Futures,* 76.

Norberg-Hodge quotation ("I have never . . . your surroundings.") is from Norberg-Hodge, *Ancient Futures,* 85.

96   Statement regarding people's universal response to the Ladakh story was made by John Page, quoted in *Local Futures: Beyond the Global Economy,* prod. ISEC, 1998, videocassette.

Dukchen Rinpoche quotation ("Your message . . . to take root.") is from Becky Tarbotton, "Notes from a Women's Alliance Village Meeting," *Local Futures* (ISEC newsletter, spring 2000): 3.

97   Ecological footprint information is in Mathis Wackernagel and William E. Rees, *Our Ecological Footprint: Reducing Human Impact on the Earth* (Gabriola Island, British Columbia: New Society Publishers, 1996); and S. Levin, ed., *Encyclopedia of Biodiversity* (San Diego, Calif.: Academic Press, 2001), s.v. "ecological footprint, concept of."

99   Norberg-Hodge quotation ("As a result . . . traditional economy.") is from Norberg-Hodge, *Ancient Futures,* 119.

Data and information on trade are in: Helena Norberg-Hodge, "Shifting Direction: From Global Dependence to Local Interdependence," (Berkeley, Calif.: ISEC, 2000); and Norberg-Hodge, Goering, and Page, *From the Ground Up,* xiii–xvi.

Daly quotation ("Americans import . . . more efficient.") is from Herman Daly, "The Perils of Free Trade," *Scientific American* 269 (1993): 51.

Information on the Bretton Woods conference and global trade organizations can be found in Norberg-Hodge, Goering, and Page, *From the Ground Up.*

100   Information on food economy and on local food is found in: Norberg-Hodge, Merrifield, and Gorelick, *Bringing the Food Economy Home;* and Brian Halweil, "Home Grown: The Case for Local Food in a Global Market," *Worldwatch Paper* 163 (Washington, D.C.: Worldwatch Institute, 2002).

101   Information on subsidies can be found in: A. de Moor and P. Calamai, *Subsidizing Unsustainable Development: Undermining the Earth with*

Public Funds (The Hague: Institute for Research on Public Expenditure; San Jose, Costa Rica: Earth Council, 1997); N. Meyers and J. Kent, *Perverse Subsidies: How Tax Dollars Can Undercut Both the Environment and the Economy* (Washington, D.C.: Island Press, 2001); and Worldwatch Institute, *Vital Signs 2003: The Trends That Are Shaping Our Future* (New York: W. W. Norton, 2003).

102   Information on the 1999 WTO protest in Seattle is found in: Paul Hawken, "Seattle: Bearing Witness," *Annals of Earth* 18 (2000): 7–9; and Norberg-Hodge, Goering, and Page, *From the Ground Up,* xi–xii.

104   The George W. Bush administration's decision in June 2002 not to heed the climate change warnings of the United States National Academy of Sciences and the United Nations Intergovernmental Panel on Climate Change, because the administration feared any change in its environmental policy would disrupt the economy, can be found in Andrew C. Revkin, "White House Shifts on Global Warming," *Times Union,* June 3, 2002, sec. D, 1, 4. For U.S. Climate Action Report (U.S. Department of State, May 2002), see http://yosemite.epa.gov/oar/globalwarming.nsf/content/ResourceCenterPublicationsUSClimate ActionReport.html.

105   The data on money spent globally on advertising are in Worldwatch Institute, *Vital Signs 2003,* 48–49.

107   Biophilia is discussed in: Edward O. Wilson, *Biophilia* (Cambridge: Harvard University Press, 1984); and Stephen R. Kellert and Edward O. Wilson, eds., *The Biophilia Hypothesis* (Washington, D.C.: Island Press/Shearwater Books, 1993).

       Advocating for the living world appears associated with early experiences, as does biophobia, considered in: David W. Orr, "Love It or Lose It: The Coming Biophilia Revolution," in *Earth in Mind: On Education, Environment, and the Human Prospect* (Washington, D.C.: Island Press, 1994), 131–153; Heather Newbold, ed., *Life Stories: World-Renowned Scientists Reflect on Their Lives and the Future of Life on Earth* (Berkeley: University of California Press, 2000).

109   Lester R. Brown's perspectives on feeding the world are in: Lester R. Brown, *Who Will Feed China?: Wake-up Call for a Small Planet* (New

York: W. W. Norton, 1995); and Lester R. Brown, *The Earth Policy Reader* (New York: W. W. Norton, 2002).

Data and information on Cuba's resolution of its food crisis in the 1990s are in Fernando Funes et al., *Sustainable Agriculture and Resistance: Transforming Food Production in Cuba* (Milford, Conn.: Food First Books, 2002).

SIX: BE FRUITFUL AND FEW

Werner Fornos quotes not identified below are from my interview with him on 25–30 March 2001.

112   Epigraph: Werner Fornos (lecture at Madison University, Va., March 28, 2001).

Werner Fornos's ideas and perspectives are presented in his book: *Gaining People, Losing Ground* (Ephrata, Penn.: Science Press, 1987); and in the many articles and news releases of the Population Institute, 107 Second Street, N.E., Washington, D.C. 20002, including *Popline: World Population News Service*, a bimonthly; http://www.population-institute.org.

113   The 65 factually based calculations that give a mean of 12 billion people are discussed by Joel E. Cohen, *How Many People Can the Earth Support?* (New York: W. W. Norton, 1995). This mean is for the calculations that give a range with a maximum and minimum. For these calculations the median low value is 8 billion and the median high value is 16 billion.

Warnings about the consequences of the growing human population are given in Paul R. Ehrlich, *The Population Bomb* (New York: Ballantine, 1968). Together with his wife, Paul Ehrlich wrote a more recent book highlighting problems of overpopulation: Paul R. Ehrlich and Anne H. Ehrlich, *The Population Explosion* (New York: Touchstone, 1991).

Three essays on human population are found in Thomas Malthus, Julian Huxley, and Frederick Osborn, *Three Essays on Population* (New

York: New American Library, Mentor Book, 1960). These essays were written respectively in 1830 (Malthus), 1955 (Huxley), and 1960 (Osborn). Malthus's first article, "Essay on the Principle of Population," was published in 1798.

President Richard Nixon's 1969 statement is from "Population and the American Future," no. 5250-002 (Washington, D.C.: U.S. Government Printing Office, 1972), 3.

114   Information on Werner Fornos's early life and quotations that refer to this period in his life are from *Current Biography* (New York: H. W. Wilson, 1993), 17–21.

120   The Mexico City policy quotation is from a White House press release: "Memorandum for the Administrator of the United States Agency for International Development, Subject: Restoration of the Mexico City Policy" (January 22, 2001); George W. Bush's stated goals for reinstating the Mexico City policy are from another White House press release: "Statement by the Press Secretary: Restoration of the Mexico City Policy" (January 22, 2001).

121   The accumulated evidence for evolution can be found in any introductory college biology textbook. See, for example, Peter H. Raven and George B. Johnson, *Biology*, 6th ed. (Boston: W. C. B. McGraw-Hill, 2002); and Teresa Audesirk, Gerald Audesirk, and Bruce E. Byers, *Biology: Life on Earth*, 6th ed. (Upper Saddle River, N.J.: Prentice-Hall, 2002).

SEVEN: LIVING IN A FINITE WORLD

Herman Daly quotes not identified below are from my interview with him on July 22–23, 2002.

132   Epigraph: Herman E. Daly, *Beyond Growth: The Economics of Sustainable Development* (Boston: Beacon Press, 1996), 145.

        Herman Daly's ideas and perspectives are presented in his books: Herman E. Daly and Kenneth N. Townsend, *Valuing the Earth: Economics, Ecology, Ethics* (Cambridge: MIT Press, 1993); Herman E. Daly and John B. Cobb, Jr., *For the Common Good: Redirecting the Economy toward the Environment and a Sustainable Future,* updated and expanded ed. (Boston: Beacon Press, 1994); Herman E. Daly, *Beyond Growth: The Economics of Sustainable Development;* and Herman E. Daly, *Ecological Economics and the Ecology of Economics: Essays in Criticism* (Northhampton, Mass.: Edward Elgar, 1999).

The physics bias of early economists is given in Daly and Cobb, *For the Common Good,* 28–30.

A history of economic thought is found in Robert L. Heilbroner, *The Worldly Philosophers: The Lives, Times, and Ideas of the Great Economic Thinkers,* 7th ed., rev. (New York: Touchstone, 1999).

133   Francis Edgeworth's assumption is from Heilbroner, *Worldly Philosophers,* 173.

134   The *Homo economicus* caricature is from Daly and Cobb, *For the Common Good,* 5, 85–96.

Discussion of measures for utility and economic health other than GDP is found in Daly and Cobb, *For the Common Good,* 62–84, 443–507.

Creating better models for economic behavior is considered in Herbert Gintis, "Beyond *Homo Economicus:* Evidence from Experimental Economics," *Ecological Economics* 35 (2000): 311–322; Herbert Gintis, *Game Theory Evolving: A Problem-Centered Interaction to Modeling Strategic Interaction* (Princeton: Princeton University Press, 2000); and Colin F. Camerer, *Behavioral Game Theory: Experiments in Strategic Interactions* (Princeton: Princeton University Press, 2003).

135   Herman Daly's criticism of *Homo economicus* is from Daly and Cobb, *For the Common Good,* 159–175.

139   Nicholas Georgescu-Roegen's understanding of entropy and economics is given in Nicholas Georgescu-Roegen, "The Entropy Law and the Economic Problem," reprinted in Daly and Townsend, *Valuing the Earth,* 75–88.

141   Herman Daly's story of the 1992 World Bank environmental report is from Daly, *Beyond Growth,* 5–7.

Daly quotation ("[T]he larger box . . . dependent on it.") is from Daly, *Beyond Growth*, 6.

Summer quotation ("That's not . . . look at it.") is from Daly, *Beyond Growth*, 6.

142   Daly quotation ("could not . . . more irreconcilable.") is from Daly, *Beyond Growth*, 7.

143   John Stuart Mill's 1857 assessment is from Daly, *Beyond Growth*, 3.

Herman Daly's ideas and concepts concerning the steady-state economy and uneconomical growth are given in Herman E. Daly, "Sustainable Development: Definitions, Principles, Policies" (World Bank address, Washington, D.C., April 30, 2002).

145   Daly's quotation ("Ecological limits . . . caused them.") is from Daly, "Sustainable Development."

The information on ocean fisheries is from Richard Ellis, *The Empty Ocean* (Washington, D.C.: Island Press, 2003), 11–92; D. Pauly et al., "Fishing Down Marine Food Webs," *Science* 279 (1998): 860–863; and Ransom A. Myers and Boris Worm, "Rapid Worldwide Depletion of Predatory Fish Communities," *Nature* 423 (2003): 280–283.

147   Daly quotation ("What good . . . aquifer or river?") is from Daly, *Beyond Growth*, 77.

A discussion of the New York City water-supply situation is given in Daly, *Beyond Growth*, 61–70; and Daly, "Sustainable Development."

148   Discussion of the Troy, New York, water system is given in Robert Moore, *A Comprehensive History of the Potable Water Supply in Troy, New York* (Troy: Robert Moore, 1991).

A discussion of market capacity for efficient allocation but not for appropriate scale is given in Daly and Cobb, *For the Common Good*, 49–61.

149   Daly quotation ("This absolute . . . biospheric ark.") is from Daly, *Beyond Growth*, 50–52.

150   A discussion of the relationship between Frederick Soddy's writings and economics is given in Daly, *Beyond Growth*, 171–190.

A discussion of money is given in Daly and Cobb, *For the Common Good,* 407–442.

152 John Kenneth Galbraith quotation ("The process . . . is repelled.") is from David Colander, *Economics,* 4th ed. (New York: McGraw-Hill, 2001), 647.

155 A discussion of trade and comparative advantage are found in Daly and Cobb, *For the Common Good,* 209–235.

158 Daly quotation ("We have here . . . 'free' trade.") is from Daly, *Beyond Growth,* 155.

Daly and Cobb quotation ("Free trade . . . mutual benefit.") is from Daly and Cobb, *For the Common Good,* 220–221.

159 The assessment that money spent on local food generates more local business than money spent on nonlocal food is from Daly and Cobb, *For the Common Good,* 367. (Cf. Sam Passmore, "Hendrix Turns to Arkansas Produce," *Arkansas Gazette,* June 10, 1987.)

A discussion of tariffs and comparative advantage is in Daly, *Ecological Economics and the Ecology of Economics,* 123–127; and Daly and Cobb, *For the Common Good,* 283–288.

The free trade of information, the Thomas Jefferson quotation ("Knowledge is . . . of [hu]mankind."), and the Daly quotation ("Once knowledge . . . monopolized item.") are found in Herman Daly, "Globalization and Its Discontents" (Aspen Institute, "Globalization and the Human Condition," Fiftieth Anniversary Conference, Aspen, Colo., August 20, 2000).

160 A discussion of the larger issues associated with genetic engineering is given in David Ehrenfeld, "Widening the Context in the Biotechnology Wars" (lecture at E. F. Schumacher Society, May 11, 2002); and David Ehrenfeld, "The Cow Tipping Point," *Harper's* 305 (October 2002), 13–20.

162 Daly quotations ("[G]rowth has . . . natural world." and "We should . . . these conditions.") are from Daly, *Beyond Growth,* 219–220.

EIGHT: ACCEPTING UNCERTAINTY

Stephen Schneider quotes not identified below are from my interview with him on June 13–15, 2002.

164   Epigraph: Stephen H. Schneider interview, July 13–15, 2001.

Stephen H. Schneider's ideas and perspectives are presented in his books: Stephen H. Schneider and Lynne Mesirow, *The Genesis Strategy: Climate and Global Survival* (New York: Plenum, 1976); Stephen H. Schneider, *Global Warming: Are We Entering the Greenhouse Century?* (San Francisco: Sierra Club Books, 1989); and Stephen H. Schneider, *Laboratory Earth: The Planetary Gamble We Can't Afford to Lose* (New York: Basic Books, 1997).

The influence of climate warming on biodiversity is given in Stephen H. Schneider and Terry L. Root, eds., *Wildlife Responses to Climate Changes: North American Case Studies* (Washington, D.C.: Island Press, 2002); A. H. Fitter and R. S. R. Fitter, "Rapid Changes in Flowering Time in British Plants," *Science* 296 (2002): 1689–1691; and Daniel Grossman, "Spring Forward," *Scientific American* 290 (2004): 84–91.

165   Svante Arrhenius predicted that a doubling of CO_2 concentration would result in a warming of 8.8°F to 11.0°F for surface temperature, depending upon latitude and season (S. Arrhenius, "On the Influence of the Carbonic Acid in the Air upon the Temperature of the Ground," *Philosophical Magazine and Journal of Science* 41 [1896]: 237–276). Interestingly, these numbers are essentially identical to the higher temperatures given by the IPCC third report in 2001, in Robert T. Watson et al., *Climate Change 2001: Synthesis Report, Summary for Policymakers,* www.ipcc.ch/ (accessed June 1, 2004).

Al Gore's assessment of the importance of environmental issues is given in Al Gore, *Earth in the Balance: Ecology and the Human Spirit* (Boston: Houghton Mifflin, 1992).

166   Will quotation ("Stephen Schneider . . . today's panic-mongers.") from George F. Will, "Al Gore's Green Guilt," *Washington Post*, Thursday, September 3, 1992, sec. A, pp. 23.

The possibility of an ice age is in S. I. Rasool and S. H. Schneider, "Atmospheric Carbon Dioxide and Aerosols: Effects of Large Increases on Global Climate," *Science* 173 (1971): 138–141.

168   Schneider quotation ("climatic theory . . . if not sooner.") is from Schneider and Mesirow, *The Genesis Strategy*, 11.

Data for Earth temperature as a result of heat-trapping gases is from Schneider, *Global Warming*, 14–16.

169   Changes in carbon dioxide concentration are from John Houghton, *Global Warming: The Complete Briefing*, 2d ed. (Cambridge: Cambridge University Press, 1994), 25; and Robert T. Watson et al., *Climate Change 2001: Synthesis Report, Summary for Policymakers*, www.ipcc.ch/ (accessed June 1, 2004).

Discussion of Schneider's response to George Will's column is found in Stephen H. Schneider, "Is the 'Scientist-Advocate' an Oxymoron?" (paper presented at the annual meeting of the American Association for the Advancement of Science, Boston, Mass., February 12, 1993).

172   The story of Stephen Schneider's move to the National Center for Atmospheric Research in Boulder, Colorado, and what happened there is, in part, from Stephen H. Schneider, "Both Sides of the Fence: The Scientist as Source and Author," chap. 16 in *Scientists and Journalists: Reporting Science as News*, ed. Sharon M. Friedman, Sharron Dunwoody, and Carol L. Rogers (New York: Free Press, 1986), 215–222.

174   Quotation ("for his ability . . . with colleagues.") stating criteria for awarding a MacArthur Fellowship to Stephen Schneider is from http://www.iis.stanford.edu/mediaguide/2227/ (accessed June 1, 2004).

Factors that influence climate can be found in Edward Bryant, *Climate Process and Change* (Cambridge: Cambridge University Press, 1997).

175   History of regional temperature changes from IPCC Third Assessment Report in D. L. Albritton et al., Climate Change 2001, The Sci-

entific Basis, Technical Summary, www.ipcc.ch/ (accessed June 1, 2004).

176 Events correlated with a warming climate are from Union of Concerned Scientists, "Global Warming: Early Warning Signs," (Cambridge, Mass.: Union of Concerned Scientists, 1999).

Information on climate modeling, changes in heat-trapping gases, and links to human activities can be found in Schneider, *Laboratory Earth;* Stephen H. Schneider, congressional testimony, July 10, 1997, http://epw.senate.gov/105th/schn0710.htm (accessed July 18, 2004); Houghton, *Global Warming;* and Watson et al., *Climate Change 2001.*

178 The IPPC 1996 Working Group I quotation ("Future unexpected, . . . ecosystem changes.") is from Schneider, *Laboratory Earth,* 90.

179 Rate of warming after the last ice age is from Schneider, *Laboratory Earth,* 52.

Projected temperature increases are from Watson et al., *Climate Change 2001.*

180 Information on William Nordhaus's DICE model is in William D. Nordhaus, "To Slow or Not to Slow: The Economics of the Greenhouse Effect," *The Economic Journal* 101 (1991): 920–937; William D. Nordhaus, "An Optimal Transition Path for Controlling Greenhouse Gases," *Science* 258 (1992): 1315–1319; and William D. Nordhaus, *Managing the Global Commons: The Economics of Climate Change* (Cambridge: MIT Press, 1994).

Prediction of significant costs to the economy if we were to move away from fossil fuels is given in W. D. Nordhaus and J. Boyer, *Warming the World: Economic Models of Global Warming* (Cambridge: MIT Press, 2000).

181 Schneider's response to the conclusion that Nordhaus drew from DICE is given in S. H. Schneider, "Pondering Greenhouse Policy," *Science* 259 (1993): 1381.

The revised response in 2002 to economists' assessment that reducing carbon dioxide emissions is too expensive is given in Christian Azar and Stephen H. Schneider, "Are the Economic Costs of Stabilizing the Atmosphere Prohibitive?" *Ecological Economics* 42 (2002): 73–80.

An in-depth consideration of the economics of climate change is given in Clive L. Spash, *Greenhouse Economics: Value and Ethics* (London: Routledge, 2002).

182 Discussion of the Gulf Stream and the consequences of its alternative patterns are from Wallace S. Broecker, "Thermohaline Circulation, the Achilles Heel of Our Climate System: Will Man-Made CO_2 Upset the Current Balance?" *Science* 278 (1997): 1582–1588; William Calvin, "The Great Climate Flip-Flop," *Atlantic Monthly,* January 1998, 47–64; Brad Lemley, "The New Ice Age," *Discover,* September 2002, 35–41; and Schneider, *Laboratory Earth,* 52.

186 Wilson quotation ("Physicists can . . . complex subject than physics.") is from Edward O. Wilson, *The Diversity of Life* (Cambridge: Harvard University Press, 1992), 181.

The use of subjective judgments to assess likelihood is considered in Schneider, *Laboratory Earth,* 114–154.

188 The conflict that scientists face when entering into public discussions and trying to influence public policy is found in Schneider, "Is the 'Scientist-Advocate' an Oxymoron?"

Schneider quotation ("The double ethical bind . . . of knowledge.") is from Schneider, *Global Warming,* xi.

190 Schneider quotation ("[B]y far the most serious environmental [problem] . . . these factors.") is from Schneider, *Laboratory Earth,* 111.

NINE: ECOLOGICAL DESIGN

David Orr quotes not identified below are from my interviews with him on May 27–29, 2000, and January 16 and 25, 2004.

191 Epigraph: David W. Orr, *The Nature of Design: Ecology, Culture, and Human Intention* (Oxford: Oxford University Press, 2002), 4.

David Orr's ideas and perspectives on environmental education and ecological design are presented in three books: *Ecological Literacy: Education and the Transition to a Postmodern World* (Albany: State University of New York Press, 1992); *Earth in Mind: On Education, Envi-*

ronment, and the Human Prospect (Washington, D.C.: Island Press, 1994); and *The Nature of Design: Ecology, Culture, and Human Intention.*

192   Data on entering first-year college students is published by Cooperative Institutional Research Program, Higher Education Research Institute, Graduate School of Education and Information Studies, University of California, Los Angeles. Most recent data for the class entering in 2003 published in December 2003. In the late 1960s about 40 percent of entering students considered as an objective for attending college "being very well off financially" as "essential" or "very important." The percentage of students for whom this objective is "essential" or "very important" increased steadily until in the mid 1980s it crossed the 70 percent level and has remained over 70 percent for two decades. In 2003 the only objective more important than "being very well off financially" (73.8 percent) was "raising a family" (74.8 percent). Over the same period, "to develop a meaningful philosophy of life" dropped from an all-time high of 85.8 percent in 1967 to an all-time low of 39.3 percent in 2003. In 2003, the three highest ranked "very important" reasons for deciding to go to college were: "To learn more about things that interest me" (76.9 percent), "To be able to get a better job" (70.1 percent), and "To get training for a specific career" (70.0 percent). These data are for all four-year institutions in the United States.

194   The history of environmental studies at Oberlin is given in Karen Schaefer, "Retiring Professor of Biology David Egloff Instrumental in Creating Environmental Studies Program at Oberlin College," *The Observer* (Oberlin College) (May 22, 1998): 9.

201   Orr and Scott Momaday quotation from David W. Orr, *Earth in Mind,* 205.

204   Orr quotations ("First, we . . . education," and "Does four . . . vandals'?") are from "What Good Is a Great College if You Don't Have a Decent Planet to Put It On?" *The Observer* (Oberlin College) (November 22, 1990): 6.

205   Orr quotation ("The design, . . . catalog.") on pedagogy of academic
      buildings from David Orr, "Architecture as Pedagogy: Environmental
      Studies 167," *The Observer* (Oberlin College) (May 13, 1993): 6.

      Details of Adam Joseph Lewis Center are given in Marci Janas,
      Oberlin Alumni Magazine (fall 1998): 28–31; "Adam Joseph Lewis Cen-
      ter for Environmental Studies: Building Performance Data" (architect
      data sheet, William McDonough + Partners, 1998); and Zoë Ingalls,
      "Green Building at Oberlin Is a New Dream House for Environmen-
      tal Studies," *The Chronicle of Higher Education* (January 12, 2000): sec.
      B, p. 2. Adam Joseph Lewis Center for Environmental Studies home
      page, http://www.oberlin.edu/ ajlc/ajlcHome.html (accessed July 18,
      2004). Information on the performance of the Lewis Center is from
      conversations with David Benzing, David Orr, and John Petersen on
      many occasions after the center opened in 1998.

208   David Orr, "2020: A Proposal," *Conservation Biology* 14 (2000): 338–
      341.

214   Orr quotation ("Throughout history . . . all life.") is from *Ecological
      Literacy,* 162.

TEN: CAN WE CHANGE, WILL WE CHANGE?

216   Epigraph: Jo Anne Van Tilburg, *Easter Island: Archaeology, Ecology,
      and Culture* (Washington, D.C.: Smithsonian Institute Press, 1994),
      163–164.

      The scientific consensus that current patterns of living cannot per-
      sist can be found in hundreds of books and an immense scientific lit-
      erature. This consensus is presented in the books by the people about
      whom I have written, in notes for the preceding chapters, and in Fur-
      ther Reading.

      Discussion of humans as the dominant evolutionary force on the
      planet is found in S. R. Palumbi, "Humans as the World's Greatest
      Evolutionary Force," *Science* 239 (2001): 1786–1790.

217   Vote on the Endangered Species Act is in Edward O.Wilson, *The
      Future of Life* (New York: Knopf), 185.

Information on Al Gore's 1988 campaign for the presidential nom-
ination is from Al Gore, *Earth in the Balance* (Boston: Houghton
Mifflin, 1992), 8–9.

219   Assessment of U.S. progress toward sustainability is presented in John
C. Dembach, ed., *Stumbling toward Sustainable Development* (Washington, D.C.: Environmental Law Institute, 2002).

Christopher Shays's talk was given at the United Nations 2002
World Summit on Sustainable Development, Johannesburg, South
Africa, in the session "Stumbling toward Sustainable Development in
the United States: Where Do We Go Next?" (August 28, 2002).

220   Frazier quotation ("Of the many new congregations . . . to be stirred.")
is from Donald Hart Frazier, "Reflections on 60 Years in the Ministry"
(Center Church, New Haven, Conn., July 6, 1996).

221   Lomborg quotation ("We are actually leaving the world . . . that is a
beautiful world.") is from Bjørn Lomborg, *The Skeptical Environmentalist: Measuring the Real State of the World* (New York: Cambridge University Press, 2001), 351–352.

Washington Post quotation ("Bjørn Lomborg's good news . . . a
magnificent achievement.") is from Dennis Dutton, "Greener Than
You Think: 'The Skeptical Environmentalist: Measuring the Real State
of the World' by Bjørn Lomborg," *Washington Post,* Sunday, October
21, 2001, pp. BW01, http://www.washington post.com/ac2/wp-dyn/
A12789-2001Oct18?language=printer, (accessed June 1, 2004).

Earlier books like *The Skeptical Environmentalist* that essentially
contradict the consensus of the scientific community concerning the
seriousness of our environmental challenges are represented by Gregg
Easterbrook, *Moment on the Earth: The Coming Age of Environmental
Optimism* (New York: Viking, 1995); Julian L. Simon, *The Ultimate
Resource* (Princeton, N.J.: Princeton University Press, 1981); and Julian
L. Simon, *The Ultimate Resource 2* (Princeton, N.J.: Princeton Univer-
sity Press, 1998).

222   Winston Churchill quotation ("A lie gets halfway around the
world . . . its pants on.") is from "Union of Concerned Scientists
Examines *The Skeptical Environmentalist,*" http://www.ucsusa.org/

global_environment/archive/page.cfm?pageID+533 (accessed June 1, 2004).

See the Union of Concerned Scientists website for information on the scientific response to Lomborg's book, http://www.ucsusa.org/ global_environment/archive/page.cfm?pageID+533 (accessed June 1, 2004), or write to them at: Union of Concerned Scientists, 2 Brattle Square, Cambridge, Mass. 02238.

Wilson quotation ("My greatest regret about the Lomborg scam . . . the slow process of peer review and approval.") is from E. O. Wilson, "Vanishing Point: On Bjørn Lomborg and Extinction," *Grist Magazine,* http://www.gristmagazine.com/ grist/books/ wilson121201.asp (accessed June 1, 2004).

The general flavor of the global scientific response to *The Skeptical Environmentalist* is given in Stuart Pimm and Jeff Harvey, "No Need to Worry about the Future: Environmentally, We Are Told, 'Things Are Getting Better,'" *Nature* 414 (2001): 149–150; Michael Grubb, "Relying on Manna from Heaven?" *Science* 294 (2001): 1285–1286; John Rennie et al., "Science Defends Itself against Misleading Math about the Earth," *Scientific American* (January 2002): 61–71.

224 Failures of human civilizations are discussed in Jared Diamond, *The Third Chimpanzee: The Evolution and Future of the Human Animal* (New York: HarperCollins, 1992); Tim Flannery, *The Future Eaters* (Port Melbourne, Victoria: Reed Books, 1994); P. Bahn and J. Flenley, *Easter Island, Earth Island* (London: Thames and Hudson, 1992); Van Tilburg, *Easter Island;* and Clive Ponting, *A Green History of the World: The Rise and Fall of Great Civilizations* (New York: Penguin Books, 1991).

225 An overview of human nature can be found in Edward O. Wilson, *On Human Nature* (Cambridge: Harvard University Press, 1978); Edward O. Wilson, *Consilience* (New York: Knopf, 1998); Reg Morrison, *The Spirit in the Gene: Humanity's Proud Illusion and the Laws of Nature* (Ithaca, N.Y.: Cornell University Press, 1999); Bobbi S. Low, *Why Sex Matters: A Darwinian Look at Human Behavior* (Princeton, N.J.:

Princeton University Press, 2000); and Steven Pinker, *The Blank Slate: The Modern Denial of Human Nature* (New York: Allen Lane, 2002).

226   Wilson quotation ("The human brain . . . crumble around them.") is from Wilson, *The Future of Life,* 40.

Discussions of successful human cultures including !Kung, Nauruan, Tikopian, and Ladakhi are given in John Gowdy, ed., *Limited Wants, Unlimited Means: A Reader on Hunter-Gatherer Economics and the Environment* (Washington, D.C.: Island Press, 1998); Helena Norberg-Hodge, *Ancient Futures: Learning from Ladakh* (San Francisco: Sierra Club Books, 1992); and Carl N. McDaniel and John M. Gowdy, *Paradise for Sale: A Parable of Nature* (Berkeley, C.A.: University of California Press, 2000).

Further Reading

Each of the eight people about whom I have written was asked to suggest half a dozen books that explore the issues he or she has championed and that would be appropriate for the general reader. The suggested readings are listed below by chapter number and subject. The list under the chapter 10 heading indicates six books I encourage everyone to read.

TWO. TERRI SWEARINGEN: *Pollution and Social Justice*

Michelle Allsopp, Pat Costner, and Paul Johnson, *Incineration and Human Health: State of Knowledge of the Impact of Waste Incinerators on Human Health* (Exeter, U.K.: Greenpeace Research Laboratories, 2001).

Keeny Ausubel, ed., *Ecological Medicine: Healing the Earth, Healing Ourselves* (San Francisco: Sierra Club Books, 2004).

Theo Colborn, Dianne Dumanoski, and John Peterson Myers, *Our Stolen Future: Are We Threatening Our Fertility, Intelligence, and Survival?—A Scientific Detective Story* (New York: Dutton, 1996).

Ken Geiser, *Materials Matter: Toward a Sustainable Materials Policy* (Cambridge: MIT Press, 2001).

Carolyn Raffensperger and Joel Tickner, eds., *Protecting Public Health and Environment: Implementing the Precautionary Principle* (Wasington D.C.: Island Press, 1999).

Sheldon Rampton and John Stauber, *Trust Us We're Experts: How Industry Manipulates Science and Gambles with Your Future* (New York: Tracher/ Putnam, 2001).

THREE. DAVE FOREMAN: *Conservation and Biodiversity*

David Maehr, Reed Noss, and Jeffery Larkin, eds., *Large Mammal Restoration* (Washington, D.C.: Island Press, 2001).

Reed Noss and Allen Cooperrider, *Saving Nature's Legacy* (Washington, D.C.: Island Press, 1994).

Michael Soulé and John Terborgh, eds., *Continental Conservation* (Washington, D.C.: Island Press, 1999).

John Terborgh, *Requiem for Nature* (Washington, D.C.: Island Press, 1999).

Edward O. Wilson, *The Diversity of Life* (Cambridge: Harvard University Press, 1992).

———. *The Future of Life* (New York: Knopf, 2002).

FOUR. WES JACKSON: *Agriculture and Place-Based Communities*

Wendell Berry, *Home Economics* (New York: North Point Press, 1987).

———. *Unsettling of America: Culture and Agriculture* (San Francisco: Sierra Club Books, 1977).

Daniel Hillel, *Out of the Earth: Civilization and the Life of Soil* (Berkeley: University of California Press, 1992).

Andrew Kimbrell, ed., *The Fatal Harvest Reader: The Tragedy of Industrial Agriculture* (Washington, D.C.: Island Press, 2002).

Judith D. Soule and Jon K. Piper, *Farming in Nature's Image: An Ecological Approach to Agriculture* (Washington, D.C.: Island Press, 1992).

William Vitek and Wes Jackson, eds., *Rooted in the Land* (New Haven: Yale University Press, 1996).

FIVE. HELENA NORBERG-HODGE: *Globalization and Living Locally*

Derrick Jensen, *The Culture of the Make Believe* (New York: Context Books, 2002).

———. *A Language Older Than Words* (New York: Context Books, 2000).

David Orr, *Earth in Mind: On Education, Environment and the Human Prospect* (Washington, D.C.: Island Press, 1994).

Majid Rahnema, ed., *The Post-Development Reader* (London: Zed Books, 1997).

E. F. Schumacher, *Small Is Beautiful: Economics as if People Mattered* (New York: Harper and Row, 1973).

Michael Schuman, *Going Local: Creating Self-Reliant Communities in a Global Age* (New York: Routledge, 2000).

SIX. WERNER FORNOS: *Population and Consumption*

Lester R. Brown, *Who Will Feed China: Wake-up Call for a Small Planet* (New York: W. W. Norton, 1995).

Joel E. Cohen, *How Many People Can the Earth Support?* (New York: Norton, 1995).

Paul R. Ehrlich and Anne H. Ehrlich, *The Population Explosion* (New York: Touchstone, 1990).

———. *One with Nineveh: Politics, Consumption, and the Human Future* (Washington, D.C.: Island Press, 2004).

Donella H. Meadows, Jorgen Randers, and Dennis L. Meadows, *Limits to Growth: The 30-Year Update* (White River Junction, Vt.: Chelsea Green Publishing, 2004).

Mathis Wackernagel and William Rees, *Our Ecological Footprint: Reducing Human Impact on the Earth* (Gabriola Island, B.C.: New Society Publishers, 1996).

SEVEN. HERMAN DALY: *Economics and Livable Communities*

Peter Barnes, *Who Owns the Sky? Our Common Assets and the Future of Capitalism* (Washington, D.C.: Island Press, 2001).

Lester Brown, *Eco-Economy: Building an Economy for the Earth* (New York: W. W. Norton, 2001).

Brian Czech, *Shoveling Fuel for a Runaway Train: Errant Economists, Shameful Spenders, and a Plan to Stop Them All* (Berkeley: University of California Press, 2000).

Paul Hawken, Amory Lovins, and L. Hunter Lovins, *Natural Capitalism: Creating the Next Industrial Revolution* (Boston: Little, Brown, 1999).

Wes Jackson, *Becoming Native to This Place* (Lexington: University Press of Kentucky, 1994).

Carl N. McDaniel and John M. Gowdy, *Paradise for Sale: A Parable of Nature* (Berkeley: University of California Press, 2000).

EIGHT. STEPHEN SCHNEIDER: *Climate Change and Scientific Uncertainty*

Ross Gelbspan, *The Heat is ON: The High Stakes Battle over Earth's Threatened Climate* (Reading, Mass.: Addison-Wesley, 1997).

Stephen H. Schneider and Terry L. Root, eds., *Wildlife Response to Climate Change: North American Case Studies* (Washington, D.C.: Island Press, 2001).

Stephen H. Schneider, Armin Rosencranz, John O. Niles, eds., *Climate Change Policy: A Survey* (Washington, D.C.: Island Press, 2002).

Richard C. J. Somerville, *The Forgiving Air: Understanding Environmental Change* (Berkeley: University of California Press, 1996).

Clive L. Spash, *Greenhouse Economics: Value and Ethics* (London: Routledge, 2002).

Spencer R. Weart, *The Discovery of Global Warming* (Cambridge: Harvard University Press, 2003).

NINE. DAVID ORR: *Environmental Education and Ecological Design*

J. Glen Gray, *Rethinking American Education: A Philosophy of Teaching and Learning* (Middletown, Conn.: Wesleyan University Press, 1984).

Amory B. Lovins et al., *Small Is Profitable: The Hidden Economic Benefits of Making Electrical Resources the Right Size* (Snowmass, Colo.: Rocky Mountain Institute, 2002).

John Tillman Lyle, *Regenerative Design for Sustainable Development* (New York: Wiley, 1996).

J. R. McNeill, *Something New under the Sun: An Environmental History of the Twentieth-Century World* (New York: W. W. Norton, 2000).

Sim Van der Ryn and Stuart Cowan, *Ecological Design* (Washington, D.C.: Island Press, 1996).

Alfred North Whitehead, *The Aims of Education and Other Essays* (New York: Macmillan, 1929).

TEN. CARL MCDANIEL: *Life Support and Cultural Change*

Thomas Berry, *The Great Work: Our Way into the Future* (New York: Bell Tower, 1999).

Lester R. Brown, *Plan B: Rescuing a Planet under Stress and a Civilization in Trouble* (New York: W. W. Norton, 2003).

Aldo Leopold, *A Sand County Almanac and Sketches Here and There* (New York: Oxford, 1949).

Reg Morrison, *The Spirit in the Gene: Humanity's Proud Illusion and the Laws of Nature* (Ithaca: Cornell University Press, 1999).

Clive Ponting, *A Green History of the World: The Rise and Collapse of Great Civilizations* (New York: Penguin Books, 1991).

Edward O. Wilson, *Consilience* (New York: Knopf, 1998).

Index

A

Abbey, Edward, 34–36, 40

abortion, 120–21

Adam Joseph Lewis Center for Environmental Studies. *See* Lewis Center

Adams, Jonathan S., 237n

aerosols, 167, 168

Agnes Scott College, 195–96, 199

agricultural appropriations legislation, 76–77

agriculture. *See also* farmers; Sunshine Farm

  history, 63–64

  industrial, 57–59, 65, 99–100, 110–11, 159

  local, 99, 106–11

  writings on, 67–68

Albritton, D. L., 252–53n

Aldrich, Ann, 23, 26

Allison, Joy, 25

American Association for the Advancement of Science meeting, 172–74

Anderson, Ray C., 32–33, 235n

Anne Anderson, et al. v. W. R. Grace & Co., et al., 20

Arrhenius, Svente, 165, 251n

atomic bomb tests, 8

Audesirk, Gerald, 247n

Audesirk, Teresa, 247n

Azar, Christian, 181, 253n

B

Babbitt, Bruce, 217

Bahn, P., 258n

Bald Mountain, 41–42

Barzun, Jacques, 230n

Becher, Anne, 234n

Bender, Marty, 78, 241n

Berry, Thomas, 230n

Berry, Wendell, 201, 204, 230n, 239n

biodiversity, negative impact of human behavior on, 44, 49–50

Bioneers Conference, 6

biophilia, 105, 107, 201, 214–15

Biophilia (Wilson), 107

birth control. *See* family planning

Black River Cafe, 193, 210

Black River Environmental Education Partnership Project, 209

Blandy, Tom, 31

Blue River, 51–52

Boyer, J., 253n

Braungart, Michael, 32, 235n

Bretton Woods conference, 99

Broecker, Wallace S., 254n

Bronowski, J., 240n

Brookes, Jay, 234n

Brower, David, 67, 201

Brown, Don, 8

Brown, Lester R., 109, 230n, 245–46n

Brown, Peter G., 154

Brown, T. C., 232n
Browner, Carol, 25
Brudzinski, Richard, 10
Bryant, Edward, 252n
Buddhism, 90
Burdett, Hal, 124
Bush, George W., 103–4, 120, 218
Butler, Tom, 237n
Byers, Bruce E., 247n
Byers-Emmerling, Melissa, 20
Byrd International Airport Rotary
    Club, 124–25

C
Cable, Charles, 232n
Cable, Sherry, 232n
Calamai, P., 244–45n
California State University at Sacra-
    mento, 61–62
Calvin, William, 254n
Camerer, Colin F., 248n
cancer, 20, 29, 81
carbon dioxide, 165, 167, 168, 176, 177,
    179, 180. *See also* heat-trapping
    gases
    reducing emissions and levels of, 180–
    82, 208
Carson, Rachel, 195
    The Sense of Wonder, 195
    Silent Spring, 8, 30, 59, 61, 139, 221,
    230n, 232n, 240n
Cartesian Economics (Soddy), 150–52
Center for Ecological Development,
    95–96
Chester, West Virginia, 6
Chihuahua, 52
Churchill, Winston, 219, 222
civil disobedience, 16–17, 19–21, 30,
    41–42
Clean Air Act, 218
Cleveland Green Building Coalition,
    209–10

climate, 68, 174–79
Clinton, Bill, 24–26, 120, 201
    administration, 217
    agricultural appropriations legislation,
    76
    1992 presidential campaign, 22–23
Clinton, Hillary, 201
Clinton-Gore Administration, 21–22,
    30
Close Up Foundation, 122
Clovis people, 44–47, 49
Cobb, John, Jr., 155, 157, 158, 200, 248n
Cohen, Joel E., 246n
Colander, David, 250n
colleges, 192, 197–98. *See also* Agnes
    Scott College; green campus move-
    ment; Meadowcreek; Oberlin
    College
Colman, Bruce, 239n
Colombia, 130
Columbia University, 169–71
Columbus, Ohio, 15, 17
Commoner, Barry, 167
consumerism, 104–5, 149–50, 192
consumption, economic growth, and
    globalization, 104–5
contraception. *See* family planning
Conway, Arkansas, 202
Crick, Francis, 60
Cuba, 109–11
Cuban Democratic Act, 109–10
culture(s), 4, 83–85
    and survival, 224–28

D
Daedalus, 2
Dalai Lama, 90
Daly, Herman E., 99, 132–63, 180, 191,
    196, 198, 212, 227, 244n, 247–49n
    For the Common Good, 155, 157, 158,
    200, 247–48n
Darwin, Charles, 197

Daschle, Tom, 218
Davidson, Eric A., 242n
Davy, Benjamin, 231n
Dawa, Angchuk, 90
debt, 151–53
Decatur, Georgia, 195–96
Dembach, John C., 257n
Development and the Environment, 141
Diamond, Jared, 243n, 258n
Diener, Robert, 14
domestication, 64
Drake, James, 74
Dubos, René, 195
Dufty, Ken, 31
Dukchen Rinpoche, 96
Dutton, Dennis, 257n
Dynamic Integrated Climate Economic
    Model (DICE), 180–81

E
Earth First!, 35, 39–43
East Elementary, 12, 29
Easterbrook, Gregg, 221, 257n
East Liverpool, Ohio. *See* Waste Tech-
    nologies Industries
East Liverpool, Ohio 33 (ELO33), 17,
    20
East Liverpool Middle School, 13–14
ecological footprinting, 79–80, 97,
    146
ecological thinking/ecological world-
    view, 4, 197, 215. *See also* Ladakh
econocentric vs. ecocentric worldview,
    228. *See also* economic worldview
economic growth, 103–4, 136, 142, 145,
    146, 155, 162
economic integration, 157–60. *See also*
    free trade
"economic man," 134, 135
economics, 198
    appropriate price for natural capital,
        147–48, 185–86

appropriate scale for macroeconomy,
    149–50
behavior and, 135–36
biosphere and, 140–43
ethics in, 161–63
globalization and, 93, 103–4
mathematics and, 132–33
opportunity costs in the macro-
    economy, 143–46
"science" of, 133–37
economic worldview, 22–23, 103–4,
    220, 223. *See also* econocentric vs.
    ecocentric worldview
economy, steady-state
    money supply in, 150–53
ecosystem(s), complexity of, 73–75
Educate America Campaign, 122, 124,
    131
education, 213–15. *See also* colleges
    environmental, 220
    program in. *See* Meadowcreek
Egloff, David, 194
Ehrenfeld, David, 250n
Ehrlich, Anne H., 246n
Ehrlich, Paul R., 118, 127, 138, 188,
    246n
    Population Bomb, 61, 113, 246n
Eirene, Vincent, 15, 16
Eisenberg, Evan, 240n
Eisley, Loren, 195
Ellis, Richard, 249n
Enck, Judy, 31
Endangered Species Act of 1973, 37,
    217
Energy Policy Act of 2003, 218
energy sources, 139–40
entropy, 139–40
environmental conferences, 210–11
environmental groups. *See* Earth First!
Environmental Protection Agency (EPA)
    Ohio office, 13–15, 19
    Region V office, 21

EPA Region V office *(continued)*
  Resource Conservation and
    Recovery Act (RCRA) permits
    from, 6, 10, 14, 19, 20
  US, 20, 23–25, 27, 29, 39, 194
environmental warnings, 2–4
evolution, 50, 83–84, 112, 185, 186, 197,
    225–26
extinct species, 44–48

F
Fall Visitors' Days, 72, 73
family planning, 114, 118–21, 124–31
farmers. *See also* agriculture; Sunshine
    Farm
  small-scale local, 57–58, 110. *See also*
    gardens and gardening
farmers' markets, 108–9
fishing industry, global, 145–46
Fitter, A. H., 251n
Fitter, R. S. R., 251n
Flannery, Tim, 45, 237n, 258n
Flenley, J., 258n
Florida, 54–55
Florida Game and Freshwater Fish
    Commission, 54
food, locally grown, 106–9
food aid, 109–10
Foreman, Dave, 34–56, 217, 235–36n,
    238n, 239n
  creative leadership positions with
    conservation venues, 38
Forest Service, U.S., 35–36, 42, 51
Fornos, Werner, 112–31
  Gaining People, Losing Ground, 129–
    30, 246n
For the Common Good (Daly & Cobb),
    155, 157, 158, 200
Frazer, Phillip, 233n
Frazier, Donald Hart, 220, 257n
free trade, 99–100, 155–60
  of knowledge, 160–61

Free Trade Area of the Americas, 105
Friends of the Earth, 67

G
Galbraith, John Kenneth, 152
gardens and gardening, 106–8, 110–11
Gardner, Gary, 240n
General Agreement on Tariffs and
    Trade (GATT), 99–101
Genesis Strategy (Schneider &
    Mesirow), 168
genetically modified organisms
    (GMOs), 160
Georgescu-Roegen, Nicholas, 139–40,
    150, 248n
Georgia, 196
Germany, 115, 116
Gibbs, Lois, 10–11
Gibbs, W. Wayt, 243n
Gila National Forest wilderness cam-
    paign, 36
Gila Wilderness Area, 51
Gingrich, Newt, 217
Gintis, Herbert, 248n
Glantz, Michael H., 239–40n
globalization, consumption, economic
    growth, and, 104–5. *See also* eco-
    nomic integration
Global Media Awards, 128
Goddard Institute, 171–72
Goering, Peter, 242n
Goldman Environmental Prize, 27
Goodland, Robert, 153
Gore, Al, 24, 166, 169, 219
  Earth in the Balance, 22, 25, 26, 165–
    66, 233n, 251n, 257n
  failures to support environmental
    causes, 23, 25, 26, 29–30, 217,
    219–20
  presidential campaigns, 21, 217
  support for environmental causes, 21–
    22, 30, 166, 217

Gorelick, Steven, 242n
Gowdy, John M., 230n, 242n, 259n
Graham, Frank, Jr., 240n
Great Plains, 70
green campus movement, 210–11
Greenpeace, 18, 25, 26, 29
Gross, Moyne, 122, 123
gross domestic product (GDP), 133–34,
    142, 180–82
Grossman, Daniel, 251n
Grubb, Michael, 258n
Gulf Stream, shift of, 182–83

H
Haight, Karen, 14
Halweil, Brian, 244n
Harr, Jonathan, 232n, 233n
Harris, Larry, 54
Harrison, Paul, 229n
Harvey, Jeff, 258n
Hawken, Paul, 230n, 245n
heat-trapping gases, 167–69, 176–78,
    208. *See also* carbon dioxide
Heilbroner, Robert L., 248n
Heinberg, Richard, 229n
Heineken Prize in Environmental
    Science, 154
Heller, Bill, 232n
Helms-Burton Act of 1996, 110
Hendrix College, 202, 210
Hightower, Jim, 233n
Homo economicus, 134, 135
Honorary Right Livelihood Award, 154
Houck, O., 235n
Houghton, John, 252n
Hubbert, M. King, 39
Hunter, Celia, 38–39
Huxley, Julian, 246–47n

I
Icarus, 2, 190
illness, 98

incinerators, 31. *See also under* Waste
    Technologies Industries
Ingalls, Zoë, 256n
Interface, 32–33
Interface Sustainability Report of 1997, 33
Intergovernmental Panel on Climate
    Change (IPCC), 166, 168, 176, 178–
    79
International Forum on Globalization,
    102
International Monetary Fund (IMF),
    99, 101
International Society for Ecology and
    Culture (ISEC), 96, 104, 108. *See
    also* Ladakh Project

J
Jackson, Dana, 57, 61
Jackson, Wes, 57–82, 109, 111, 159, 201,
    202, 212
  writings, 239n, 241n
    Man and the Environment, 62
    New Roots for Agriculture, 67–68,
      239n
    Not Man Apart, 76
    "The Search for a Sustainable Agri-
      culture," 67
James Madison University, 125, 126
Janas, Marci, 256n
Jastrow, Bob, 171, 172
Jefferson, Thomas, 160
Jenny, Hans, 73
Johns, Cheryl, 235n
Johnson, George B., 247n
Johnston, Sadhu, 210

K
Kansas, 60
Kaufman, Hugh, 14, 20
Kellert, Stephen R., 245n
Kellogg, Will, 172
Kennedy, Edward, 218

Kent, J., 245n
Kezar, Ron, 34–35
Kidd, Sue, 242n
Kimbrell, Andrew, 240n
Koehler, Bart, 35, 41
Kotok, Sara, 210
Krulwich, Robert, 193
Kutner, Lynn S., 237n

L

Ladakh, 85–96, 98, 101, 102
Ladakh Ecological Development
    Group, 95
Ladakh Project, 95, 96
Lafayette Park, 25
Land Institute, 57, 62, 63, 70–77, 79–
    81, 106
  as embodying Jackson's work, 82
  Great Plains Research Project, 75
  Natural Systems Agriculture Advisory
    Board, 77
  Natural Systems Agriculture program,
    76–77
  Rural Communities Studies program,
    80
Lane, Mary Beth, 233n
language, 84–85
Lanting, Frans, 243n
Leh, 90, 92, 94–96, 98, 99, 106
Lemley, Brad, 254n
Leopold, Aldo, 51–53, 230n, 238n
Leutze, Emanuel Gottlieb, 43
Levin, S., 242n, 244n
Lewis, Adam Joseph, 205
Lewis, Jerry, 120
Lewis Center, 192–93, 205–8, 210–12
Likens, Gene, 201
Lomborg, Bjørn, 220–23, 257n
Lorbiecki, Marybeth, 231n, 238n
Louisiana State University (LSU), 138, 153
Love Canal, 10, 11
Lovins, Amory, 201, 230n

Lovins, L. Hunter, 230n
Low, Bobbi S., 243n, 258–59n

M

Mahon, Caroline, 242n
Maine Woods National Park, 55
Malthus, Thomas, 113, 246–47n
Maryland, 117–18
Masi, Brad, 209
Matfield Green, 79, 80
Mathematical Psychics (Edgeworth), 133
McCawley, Michael, 14
McDaniel, Carl N., 230n, 242n, 259n
McDonough, William, 32, 235n
McGinty, Katie, 23
McHarg, Ian, 195, 196
McNeill, J. R., 229n
Meadowcreek, 198–200, 202–3, 208,
    209, 211, 215
Meadows, Donella, 196, 201
medicine, traditional and Western, 98
Merrifield, Todd, 242n
Mesirow, Lynne, 168, 251n
MEXFAM, 129
Mexico City policy, 120, 216–17
Meyers, N., 245n
microeconomics, 143–45
Mill, John Stuart, 143
Moab, Utah, 39
Momaday, Scott, 201
monocultures, 69–70, 75
Moor, A. de, 244–45n
Moore, Les, 42
Moore, Robert, 249n
Morgan, Susan, 35
Morrison, Reg, 243n
Moyers, Bill, 220
Muir, John, 44
Myers, Ransom A., 249n

N

Nader, Ralph, 196

NASA's Goddard Institute, 171–72

National Academy of Sciences (NAS), 176, 189

National Center for Atmospheric Research (NCAR), 172–74

National Press Club, 122, 123

National Renewable Energy Laboratory (NREL), 207

natural capital, appropriate price for, 147–48

natural systems agriculture, 76–77

nature, awakening to the value of, 105

Nature Conservancy, 38, 48, 196

Newbold, Heather, 245n

New Hampshire, 99

9/11, 103

Nixon, Richard M., 113

Norberg-Hodge, Helena, 83–111, 159, 227
  Ancient Futures, 96, 242–44n, 259n

Nordhaus, William D., 180, 181, 253n

North Carolina, University of, 196–99

Noss, Reed, 43

nuclear bomb tests. *See* atomic bomb tests

O

Oberlin College, 192–94, 203–13
  Environmental Studies Program, 192, 194, 203–5, 209

Oberlin Student Cooperative Association (OSCA), 209

Oberlin Sustainable Agriculture Project, 209

oceanic ecosystems, 145–46

Odum, Eugene, 196

Ohio Hazardous Waste Facilities Approval Board, 10

oil industry, 39

opportunity costs in the macroeconomy, 144–46

Oregon National Resources Council, 41–42

organic farming, 59, 110

organic gardening, 108

Orr, David W., 191–214
  "What Good Is a Great College if You Don't Have a Decent Planet to Put It On," 204
  writings, 212–13, 254–56n
    Earth in Mind, 201, 213, 245n, 254–55n

Orr, Wilson, 196, 203

Osborn, Frederick, 246–47n

P

Page, John, 242n

Paldan, Gyelong, 94

Palumbi, S. R., 243n, 256n

parks, national, 55

Pauly, D., 249n

Payne, Mayor, 12

Pearce, Fred, 229n

Pelosi, Nancy, 218

perennial grasses and forbs, 72

perennializing crops, 76

pesticides, 59, 108, 110

Petersen, John, 207

Pimentel, David, 240n

Pimm, Stuart L., 74–75, 229n, 240n, 241n, 258n

Pincate Desert, 34, 39

Pincate Peak, 34–35

Pinhook Swamp, 54

Pinker, Steven, 243n

Piper, Jon K., 239n

plant architectures and associations, 69–70, 75

plant growth patterns, 66–67

Plotkin, Jeremy, 241n

polycultures, 69–70, 75

Ponting, Clive, 229n, 240n, 258n

POPLINE, 124

Population Action Council, 118–19

Population Bomb, The (Ehrlich), 113, 126

Population Institute, 118–20, 122–24, 127, 129, 131

Power, Thomas Michael, 238n

prairie agriculture, ecologically based, 70–74

Prairie Festivals, 72, 73, 76, 81

prairies, 68–70, 80–81

predators, loss of, 46–47

Pregnancy Advisory Service, 118. *See also* Population Institute

protests, 102, 169–70. *See also* civil disobedience; Waste Technologies Industries, protesting against

Q

question, wrong, 13–14, 121–22

R

Rasool, Ichtiaque, 166–68, 171, 172

Rasool, S. I., 252n

Raven, Peter H., 188, 247n

Reagan, Ronald, 39, 120, 194, 216

Rees, William E., 242n, 244n

Reilly, William, 23, 233n

Rennie, John, 258n

Rensselaer Polytechnic Institute, 209

resources, natural. *See* natural capital

RESTORE, 55

Revkin, Andrew C., 245n

rewilding, 51, 53–55

Rhein-Main air base, 116–17

Ricardo, David, 155, 156

Rice University, 137–38

Riebsame, William E., 239n

Roadless Area Review and Evaluation (RARE and RARE II), 35–36, 42

Roberts, Jane, 119–20

Roberts, Walter, 174

Rockefeller, John D., 113

Rockefeller Commission, 113–14, 120, 216

Root, Terry L., 251n

Roots of Change, 104

Roselle, Mike, 42

Rotary clubs, 124–26, 128–29

rural vs. urban populations, 105–6

Rutledge, Mike, 233n

S

Sabel, Naomi, 210

Sacramento, California, 61–62

Sagan, Carl, 22, 189

sagebush rebellion, 39

San Rafael, California, 6

Schaefer, Karen, 255n

Schneider, Stephen H., 164–90, 201, 251–54n

Genesis Strategy, 168

Schregardus, Donald, 19

Schumacher, E. F., 198

Small Is Beautiful, 63, 93, 196, 230n, 240n

Schwab, Jim, 231n

Scott, Agnes, 196

Sease, Debbie, 36, 39

Seattle, Washington, 102–3

self-interest, 162

Setterberg, Fred, 232n

Shavelson, Lenny, 232n

Shaw, Rodney, 118, 119

Shays, Christopher, 219, 220

Sheen, Martin, 15–17

Sierra Club, 41

Sierra Madre, 52

Simon, Julian L., 126, 221, 257n

Simon (atomic bomb test), 8

Skeptical Environmentalist, The (Lomborg), 220–23

Skinner, Walter J., 20

Sky Islands, 51, 52

Sky Islands Wildlands Network (SIWN), 51, 53

Smith, Adam, 161–62

Soddy, Frederick, 150–52

soil erosion, 58–59, 65

Soule, Judith D., 239n
Soulé, Michael E., 43, 44, 237n
Soviet Union, breakup of, 109
Spash, Clive L., 254n
Spencer, Alonzo, 24
spray-dryer systems and technology, 14–15, 19
stationary-state economy, 143
steady-state economy, 143, 146, 150, 153
Stein, Bruce A., 237n
Stein, Michael, 14
subsidies, 101, 110
Sullivan, Walter, 173
Summers, Lawrence H., 141, 143
Sunshine Farm, 77–78
sustainable agriculture, 76, 79
Svante, Arrhenius, 165
Swearingen, Terri, 5–33, 217, 231n, 234n
Sweden, 85

T
Tainter, Joseph, 230n
Tapper, Jake, 231n
Tarbotton, Becky, 244n
tariffs, 159–60. *See also* free trade
Taylor, Ronald A., 234n
Temple, Allen, 170–71
Tepfer, David, 241n
Terborgh, John, 237n
thermodynamics, laws of, 139–40
"throughput" (energy), 140
Tobin, Rebecca, 24
Todd, John, 71, 241n
Todd, Nancy Jack, 241n
Tompkins, Doug, 43
Toronto, Canada, 79–80
Torricelli bill, 109–10
Townsend, Kenneth N., 247n
toxins. *See* pesticides; waste management
trade, global, 99–103, 105
  comparative advantage as absolute advantage in present-day, 153–60

Tri-State Environmental Council (T-SEC), 13, 18, 24–27, 29
Troy, New York, 1, 8, 108, 209
Truman, Harry S., 132
Turner, Ted, 128
Turner Broadcasting, 128

U
United Nations (UN), 99, 104, 129, 166, 218
United States, indifference toward environment, 217–19
University of North Carolina at Chapel Hill, 196–99
urban vs. rural populations, 105–6

V
Van Tassel, David, 241n
Van Tilburg, Jo Anne, 256n
Village Contraceptive Empowerment Program, 129
Voinovich, George V., 15–19

W
Waal, Frans De, 243n
Wackernagel, Mathis, 242n, 244n
Waltzer, Joseph, 210
warnings, 2–3
waste management, 27–28, 32–33. *See also* Waste Technologies Industries
Waste Technologies Industries (WTI), 6, 10–17, 19, 21, 23–24, 26–31
  incinerators, 6–15, 19, 21, 24–26, 28, 30, 31
  local support for, 22
  opposition groups, 10
  protesting against, 15–19, 24–27, 29–31
  RCRA permit, 6, 10, 14, 19, 20
  test burn, 21–24
  Voinovich and, 18
Watson, James, 60

Watson, Robert T., 251n
Watt, James, 39
Wealth of Nations (Smith), 161–62
We Care network, 123–24
West Virginia, 10
West Virginia capital, 25
White House, 24–25
Wild, Alan, 239n
Wilderness Act of 1964, 37
Wilderness Society, 35–39
Wildlands Project, 43–44, 50–51, 53–56
Will, George F., 166, 169, 252n
Wilson, Edward O., 44, 55–56, 107,
    186, 212, 214, 222, 226, 229n, 230n,
    237n, 238n, 242n, 243n, 245n, 254n,
    256n, 258n
Wilson, Harlen, 194
Woburn, Massachusetts, 20

Wolke, Howie, 35, 235–36n
Women's Alliance, 96
Woodwell, George, 201
World Bank, 99, 101, 141–42, 153–54
World Population Awareness Week,
    127–28
World Population Conference, 128
World Summit on Sustainable Develop-
    ment (WSSD), 218–19
World Trade Organization (WTO),
    100–102, 105, 145, 157, 161
World War II, 115–16
Worm, Boris, 249n

Z
Zahniser, Howard, 37
Zakin, Susan, 236n

ABOUT THE AUTHOR

Carl N. McDaniel is Professor of Biology at Rensselaer Polytechnic Institute in Troy, New York. In addition, he was the founding director of the undergraduate environmental science degree program at Rensselaer. Early on, he studied insect and then plant development, but more recently his scholarly interests have focused on the interface between biology and economics. He coauthored, with economist John M. Gowdy, *Paradise for Sale: A Parable of Nature* (2000).